PRINCIPLES OF

Tsawalk

Highlighted Parts are important, don't barrier taking
Notes.

PRINCIPLES OF
Tsawalk

An INDIGENOUS APPROACH *to* GLOBAL CRISIS

Umeek

E. RICHARD ATLEO

UBCPress · Vancouver · Toronto

20 19 18 17 16 15 14 13 12 11   5 4 3 2 1

Printed in Canada on FSC-certified ancient-forest-free paper
(100% post-consumer recycled) that is processed chlorine- and acid-free.

Library and Archives Canada Cataloguing in Publication

Atleo, Eugene Richard, 1939-
Principles of tsawalk : an indigenous approach to global crisis /
Umeek (E. Richard Atleo).

Includes bibliographical references and index.
Also issued in electronic format.
ISBN 978-0-7748-2126-1 (bound); ISBN 978-0-7748-2127-8 (pbk.)

1. Nootka Indians – Religion. 2. Indian philosophy – British Columbia.
3. Traditional ecological knowledge – British Columbia. I. Title.

E99.N85A83 2011   299.7'8955013   C2011-906120-1

Canadä

UBC Press gratefully acknowledges the financial support for our
publishing program of the Government of Canada (through the Canada Book Fund),
the Canada Council for the Arts, and the British Columbia Arts Council.

This book has been published with the help of a grant from the Canadian
Federation for the Humanities and Social Sciences, through the Aid to Scholarly
Publications Program, using funds provided by the Social Sciences and Humanities
Research Council of Canada, and with the help of the K.D. Srivastava Fund.

UBC Press
The University of British Columbia
2029 West Mall
Vancouver, BC V6T 1Z2
www.ubcpress.ca

Yaa?akmis

Yaa?akmis is a word
Kindled by the explosion of creation
Meaning love
And pain
Yaa?akmis is *Qua*
That which is
Tsawalk ... One

Contents

Preface

In a view of reality described as *tsawalk* (one), relationships are *qua* (that which is). The ancient Nuu-chah-nulth assumed an interrelationship between all life forms – humans, plants, and animals. *Relationships are.* Accordingly, social, political, economic, constitutional, environmental, and philosophical issues can be addressed under the single theme of interrelationships, across all dimensions of reality – the material and the non-material, the visible and the invisible. As a consequence, certain words in the text, such as "polarity," "spiritual," "numinous," and "belief" are placed within the view of reality described as *tsawalk* – one. These definitions offer a Nuu-chah-nulth perspective on the nature of reality in that all questions of existence, being, and knowing, regardless of seeming contradictions, are considered to be *tsawalk* – one and inseparable. They are interrelated and interconnected.

This belief system, which served to integrate all dimensions of reality, is reflected in the Nuu-chah-nulth language. The root word *qua* is an example. It is used in the name of the Creator, Kʷaaʔuuc, the first part of which means "that which is" and the second part of which means "Owner of." The same root word is also used in everyday language in the saying *qʷaasasa iš,* which means "that's just the way it is" or "that's just the way s/he is," depending on the context.

The Ahousaht Dialect of the Nuu-chah-nulth Language

Pronunciation of words will vary from one dialect to another, which adds to both spelling and translation difficulties. The following word list offers an idea of approximate pronunciation of the Ahousaht dialect and indicates

the meanings of certain words. With some exceptions, the Nuu-chah-nulth
words used in this book are based on the International Phonetic Alphabet
(IPA), which accommodates the approximately forty-three letters or
sounds found in the Nuu-chah-nulth language. The primary source for the
spellings of Nuu-chah-nulth words is a dictionary published in 1991 by the
Nuuchahnulth Tribal Council (*Our World, Our Ways: T'aat'aaqsapa Cultural Dictionary,* James V. Powell, ed.). Some words, like *tsawalk*, "Nuu-
chah-nulth," and *qua*, retain the spellings used in my first book (*Tsawalk: A
Nuu-chah-nulth Worldview,* 2004), but others use the IPA ("Qua-ootz," for
example, is spelled "Ḱ\u02b7aaʔuuc"). Another source for spelling is *A Concise
Dictionary of the Nuuchahnulth Language of Vancouver Island* (Lewiston,
NY: Edwin Mellen Press, 2005), compiled and edited by John Stonham. An
additional pronunciation guide can be found on the Internet by googling the
phrase "Nuuchahnulth language."

INTERNATIONAL PHONETIC ALPHABET	PRONUNCIATION AND MEANING
Ałmaquuʔas	Aulth-ma-koo-us: a name used by the people of Ahous for the giant woman who lived in the local mountains. Her body was discovered to consist of gum or pitch, thus her name may be translated as Pitch Person or Pitch Woman.
ciḥaa	chi-haw: a word that refers to anything strange and inevitably connected to spirit
haaḥuupa	Ha-huup-a: teachings
haḥuułi	Ha-hoolth-ee: land (and its resources) owned by a chief
hamipšiƛ	Ha-mip-shitl: to recognize (in a relational manner)
ḥaw'ił	ha-wilth: chief or wealthy one in both a material and a spiritual sense
Ḥaw'iłume	Hawilth-oomee: Wealthy Mother Earth (where "wealth" includes the material and the non-material)
himwiča	him-wits-a: storytelling

hinaayiɫ ḥaw'iɫ	hee-nii-yilth hawilth: where the first word means "a place above" and the second word means "chief," which is then a reference to the Creator. The phrase combines a metaphor (a place above) with a title (chief).
hinkiic	hinkeets: chief's dance with a wolf headdress, literally a "gift-bearing" dance
iisʔaḱ	ee-sok: sacred respect
Ḱʷatyat	Kwhat-yaught: supernatural being
Ḱʷaaʔuuc	Qua-ootz: Owner of All – Creator
kʷatyiik	kwhat-yeek: heavy (in Ahousaht dialect)
Łučhaa – łučhaa	Thluch-ha: a custom that involves members of one family, on behalf of their son, going to another family to ask for a woman in marriage
Mamaɫn'i	Ma-multh-ni: originally referred to the first arrivals from Europe, who came in sailing boats that seemed to have houses on them. This word has since come to mean anyone with fair skin.
Naas and naas	"naas" means day (light) and is also used as a reference to the Creator (Naas) during prayer. "Naas" is both a name (for the Creator) and a metaphor derived from the word "day."
ʔaapḥii	Awp-haii: to be friendly
ʕintḥtinm'it	Aint-tin-mit: Son of Mucus
ʕinaak	Aii-nawk: a nurturing, cooing sound
ʔuuštaqyu	Oosh-duk-yu: shaman, doctor, completed
ʔuusumč	Oo-sum-ich: translated as "vision quest," although the root "ʔuu" carries with it the notion "to be careful"
ƛaaqišpiiɫ	Klaq-ish-piilth: one of three names given to the house of Keesta. "Klag" means liquid oil, and liquid oil was a sign of wealth, both material and spiritual.

ƛuułḥapii	Tlooth-hup-ee: go slowly
ƛuukʷaana	Tloo-qua-nah: a word that literally means "we remember reality." Anthropologists describe it as a sacred wolf ritual, which is like describing the Christian sacrament as a sacred bread ritual. The intent of this remembrance ceremony is to regain the balance and harmony of relationships between family members and thus to reflect ancient teachings.
quuʔas	Koo-us: today means "human" but originally referred to any life form
quʔac̓aqstum	Ko-ats-uk-stum: that which is (alive) in a person
Quʔušinm'it	Qu-ooshin-mit: Son of Raven
kʷist	kwist: to change, to turn (the page)
qʷaasasa ał	qua-sa-sa-ulth: just the way they are
qʷaasasa iš	qua-sa-sa-ish: just the way (he, she, or it) is
qʷaasasa sqʷi	qua-sa-sa-sqwee: that's the way it was
qʷaasasa uƛ	qua-sa-sa-ukl: that's the way it will be
t'apswiis	Tups-weese: to dive into the ocean
titiic̓u	Ti-teach-tsu: the life principle within a person
usma	usma: precious
wiiša?	Wee-shah: ceremonially unclean
wikiiš čaʔmiiḥta	Wik-eesh Cha-miih-ta: imbalance or disharmony
wikiičiƛ	wee-kee-chitl: suspend customs (momentarily)
witwok	wit-wok: a security or police force
yaaʔakmis	yaw-uk-miss: love (which includes the root word for pain)
Yukʷat	Yew-quaht: place of wind
yuxyiik	Yux-yek: meanning "heavy" in Muwachat dialect

Acknowledgments

In 1990, when I graduated from the University of British Columbia with my doctoral degree, I was asked: Who is your role model? Who influenced you and encouraged you? Ultimately, the best answer points to my ancestral legacy, my grandparents, who influenced the formative years of my life. During a retreat on Gabriola Island just across from Nanaimo, British Columbia, I performed a ceremony to honour my father, Eugene, and my grandparents, including Keesta, my father's grandfather, all of whom embody Nuu-chah-nulth cultural lifeways. My first book, *Tsawalk: A Nuu-chah-nulth Worldview,* and my current work both derive from my ancestral legacy. Innumerable others, dating back to the time of Son of Raven, are included in this legacy, so to enter each name would fill several large volumes.

In addition, my partner and colleague, Marlene, particularly during our breakfast sessions, has made important contributions to the development of the ideas used in this work. My thanks to her and to several others with whom I have consulted: my aunt Trudy and her husband, Edwin; my relative Barney Williams from Tla-o-qui-aht; my long-time friend and scholar Dr. Nancy Turner from the University of Victoria; my new friend Graham Saayman, with whom I have had, and continue to have, several discussions concerning indigenous issues of mutual concern in South Africa and Canada; as well as to my brother and illustrator, ƛiismiik. Early drafts were edited by Robert Lewis and sympathetic feedback was provided by colleagues from the University of Manitoba – Yatta Kanu, Jon Young, and John Wiens.

The shape of the final manuscript could not have come about without the very able critique of the peer review readers, the input of the Publications Board at UBC Press, and the astute mediation of Darcy Cullen. To all I extend my heart-felt gratitude.

PRINCIPLES OF
Tsawalk

Introduction

Haw'iɬume, Wealthy Mother Earth, the home of biodiversity, is currently under abnormal duress. Her immediate problem is a global warming that has produced a "dis-ease" evident in her convulsions in the form of violent storms and earthquakes. Other threats to Haw'iɬume and her inhabitants include, but are not limited to, a looming energy crisis, rampant diseases, the possibility of nuclear war, and terrorism. What has gone wrong?

It doesn't seem that long ago when certain people spread throughout the world carrying the good news about civilization and the promise of a better life. In 1860, during the time of my great-great-grandfather Haw'iɬ Nucmiis, it was the Englishman Gilbert Sproat who, out on the westernmost edge of Canada, promised Haw'iɬ Shewish of the Tseshaht peoples: "The white man will give you work." To which Shewish replied: "Ah, but we don't care to do as the white men wish ... We wish to live as we are."[1]

The interpretation of what happened after this typical encounter has now become a global issue. On the side of Gilbert Sproat are those who admire contemporary civilization for its many astonishing accomplishments in numerous fields, including medicine, communications, transportation, and technology. On the side of Haw'iɬ Shewish are the growing number of scholars concerned that Haw'iɬume, Wealthy Mother Earth, is showing distressing signs of ill health. These may yet prove, in some currently unknown way,[2] to be the dark side of civilization's astonishing accomplishments.

These astonishing accomplishments are based on proven methods, which are not in dispute. Even the scientific origin story, which can be simplified into three steps – bang,[3] evolution, and here we are – cannot, from a Nuu-chah-nulth perspective, be an issue. This is because, typically, indigenous peoples are allowed to have their own individual perspective on creation and the nature of reality. To the traditionally oriented Nuu-chah-nulth, different perspectives on creation are not a source of disagreement, confusion, or conflict; rather, they are a source of enrichment. Faced with an incomprehensible and mysterious creation, the ancient Nuu-chah-nulth came to believe that their ability to comprehend it, both ontologically and epistemologically, was so comparatively insignificant as to make hegemony a concept with no basis in reality. Who could begin to pretend to know and understand creation? Even the most powerful, the most gifted were perceived within a context that assumed their insignificance. One of many consequences of this ancient view of reality is that each person and each family were free to experience for themselves the nature of creation without being subjected to hegemonic coercion. For example, my grandmother Mary Little said emphatically: "Our stories are true!" This even though each had its own version, its own view, and its own perspective. Yet, each version of a story proved to be a reliable guide to living because, from the beginning, the method of testing the truth of each story generally consistently resulted in all basic needs (such as food, clothing, housing, and a measure of security) being met.

At this juncture of global history Ḥaw'iłume and most of her inhabitants do not enjoy the better life that was promised at the onset of colonial rule. The very foundations of the promise of civilization appear to be crumbling. Consequently, there is a suspicion that the main origin story provided by the dominant peoples of the globe, who have prevailed for the past five hundred years, may be contributing to our current global crisis. Do contemporary thinkers and scholars say with confidence, as my Granny Mary said of Nuu-chah-nulth stories, that the foundations of civilization are reliable? Some scholars do but others do not. The main storyline attributed to Charles Darwin has many variations, ranging from the purely biological to the socio-cultural to the evolution of consciousness; however, from its very inception the theory of evolution has been controversial.

For the Nuu-chah-nulth and other indigenous peoples around the world the full title to Darwin's work – *On the Origin of Species by Means of Natural Selection, or the Preservation of Favoured Races in the Struggle for Life* – had unfortunate implications. All the more so because Darwin's ideas remain influential to this day. It is often said that Darwin is misunderstood, but history is clear regarding the interpretation given to the phrase "favoured races": those of European origin were placed over those of non-European origin. Based on this storyline, it was necessary, in the words of one missionary in 1632, "to wean [indigenous peoples] from the habits and thoughts of their ancestors" and to replace these "with the acquirements of the language, arts, and customs of civilized life."[4] Regardless of whether or not Darwin's theory is misunderstood, in part or in whole, there is no question that the notion of "favoured races" translated into European hegemony over the rest of the world's peoples.

Unknown to Darwin, while he was busy developing his theory, sophisticated human societies had already evolved in various parts of the world – for example, in both Africa[5] and the Americas – centuries before the emergence of the general practice and culture of science. Charles C. Mann's 2005 publication *1491: New Revelations of the Americas before Columbus* argues:

> Today we know that technologically sophisticated societies arose in Peru first – the starting date, to archaeologists' surprise, keeps getting pushed back. Between 3200 and 2500 BC, large scale public settlements on the Peruvian coast – an extraordinary efflorescence for that time and place. When the people of the Norte Chico were building these cities, there was only one other urban complex on earth: Sumer.[6]

Had Darwin been open to the possibility of the existence of sophisticated human societies other than those found in Europe he may have been less inclined to use the phrase "favoured races" in the title of his book. The fact that an awareness that "technologically sophisticated societies arose" in the Americas in advance of European development has only recently come to light means that indigenous lifeways and knowledge systems *remain relatively unknown and unacknowledged.* Instead, indigenous peoples have been better known through such stereotypes as the Noble Savage, which, Mann explains, dates back to the 1530s, when it is found in the writing of Bartolomé de Las Casas.[7] However, in my view, the widely read letters of the explorer Amerigo Vespucci, a contemporary of Christopher Columbus, likely did more to excite the European imagination regarding the indigenous peoples of the Americas than anything else.[8] These letters may have influenced Enlightenment thinkers like Jean-Jacques Rousseau, whose work continues to be a subject of study. He imagined that the peoples of the Americas were instinctively childlike. As though to emphasize the continuing force of Rousseau's ideas on this subject, Mann notes: "In our day, beliefs about Indians' inherent simplicity and innocence refer mainly to their putative lack of impact on the environment."[9]

Since I have a culturally specific perspective on traditional Nuu-chah-nulth lifeways, which is derived from lived experience and is completely unlike the perspective that resulted in the negative stereotypical views of indigenous peoples imagined during the colonial era, I have to assume that the latter views are distortions. If they are not distortions, then how did my people come to know how to live in such a way as to inspire my Granny Mary to say with confidence: "Our stories are true!"? Not only did the Nuu-chah-nulth way of life provide every community member with the basic necessities of food, shelter, and clothing, but it also provided a rich cultural tradition of winter ceremonies that included stories, songs, oratory, exchanges of different kinds of gifts, dances, and feasting (some of which continue, in modified form, to the present day). How could ancient Nuu-chah-nulth know that, underlying the apparent inertness of much of physical reality, there is a dynamic, living world? How did they come to live, albeit not always successfully, as though personal and community well-being is dependent on, and must be inclusive of, all reality, including water, land, plants, animals, humans, and, indeed, anything that seems to be alive? How did they come to live as though there is a greater reality *beyond the surface of things?* Why is it that they considered empirically knowable reality to be an incomplete, sometimes even an unreliable, source of information?

These are difficult questions, and any answers that are contained in this book must be considered emergent and incomplete, which is to say *qʷaasasa iš,* meaning "just the way it is" (in this case referring to the nature of the human condition). To the ancient Nuu-chah-nulth, and within the *ƛakišpiƚ* (big house) into which I was born, to say *qʷaasasa iš* was not to affect a false sense of humility but to acknowledge the natural place of the human within a universe characterized primarily by mystery, which, in turn, posed a necessary challenge to growth and maturation. The natural place of the *quuʔas* (human) within a universe characterized primarily by mystery formed a significant part of the personal identity of ancient Nuu-chah-nulth. Mystery is needed to challenge the life-long process of growth and development – in body, soul (character), and spirit.

A glimpse into how the ancient Nuu-chah-nulth may have learned to negotiate this mystery is found in my first book, *Tsawalk: A Nuu-chah-nulth Worldview*. This is, necessarily, only a glimpse since the topic is the nature of reality, most of which remains unclear, obscure, and unknown, in spite of the fact that ideas about human knowledge systems fill large libraries. In *Tsawalk* I use the story of Son of Raven and his quest for the light to illustrate indigenous story-as-theory and its method of knowledge acquisition. This work expands on that glimpse into ancient Nuu-chah-nulth life and elaborates on story as theory and vision quest as method. Origin stories that serve the purpose of theory can illustrate, or model, the nature of reality. "Model," in this usage, is like a parable that can illustrate certain truths about life, requiring reflection in order to penetrate various levels of meaning. For example, in the story of Son of Raven, it is not immediately clear that it contains clues regarding how to appropriately perform a *ʔuusumč* (vision quest). Nor is it clear that success in the venture to capture the light can be interpreted as a natural order of creation, or *qua*, which implies intent. The character who owns the light is the Head Wolf among a sacred community of wolves. One Nuu-chah-nulth name for the Creator is Ḱʷaaʔuuc, Owner of All. Thus, it may be deduced that the Head Wolf in the story is Ḱʷaaʔuuc.

Gradually, from the time of beginnings, from the time of Son of Raven, over millennia, after a long and difficult struggle, the Nuu-chah-nulth developed a way of life that served, for the most part, to meet all their basic needs. This ancient way of life lacked the marvels of contemporary technology (such as the computer and other advanced methods of communication and transportation), all of which are made possible by that powerful tool of science known as the reductionist method, whereby reality, for purposes of investigation, is reduced to one or two variables. However, in the context of today's global crisis, an examination of the ancient Nuu-chah-nulth way of life may have something useful to contribute. The ideas and practices associated with it may not have been clearly understood or always successfully practised, but they can now begin to be articulated within a general framework that I discuss under several rubrics (such as constitutionalism, philosophy, worldview, political science, and even psychology), all of which come together under the general heading "Principles of *Tsawalk*."

With the onset of colonization, set off in 1492 by Columbus's search for a trade route to the East, the Nuu-chah-nulth, along with other indigenous peoples, experienced what Nigerian author Chinua Achebe describes in his novel *Things Fall Apart*.[10] Eventually, everywhere indigenous ways of life fell apart. The colonial project, in line with evolutionary theory, intended to create a better way of life in the world. The Englishman, Gilbert Sproat, indicated as much in 1860 when speaking to the Tseshaht chief, Shewish.

Today, the experience of things falling apart has become a global phenomenon, particularly with respect to two crises: (1) humanity's relationship with humanity and (2) humanity's relationship with the environment. The global reaction to the latter crisis has been identified by environmentalist-author Paul Hawken as "the largest movement in the world."[11] It is a movement in which ordinary people are not waiting for government or state action, nor are they waiting for philosophical answers. As more and more peoples witness their homelands threatened or destroyed (as is currently happening in the Republic of Maldives and in the village of Shishmaref, Alaska), they cannot afford to wait any longer.

Recently, during the aftermath of the 2004 tsunami that hit Asia, a news cameraperson filmed an Indonesian woman as she wailed: "What did we do wrong?" This cry of despair carries within it the assumption that humans and nature are intimately linked. It is as though there is one whole, within which the action of one part (the human) is related to the reaction of the other part (nature). Because an intimate connection is assumed and understood in this woman's cry, we hear no disconnect between the power of the tsunami and the living of her daily life. There is no assumption that what is happening is the result of some random process; rather, all things and all events are understood to be interrelated and interconnected.

Did the Nuu-chah-nulth peoples believe in the same kind of connections between human activities and natural phenomena as those clearly indicated by the cry of the Indonesian woman? I called my Aunt Trudy to talk about it. "Yes," she said, "we used to have songs to sing to the earth." That was enough for me. There remains today an indigenous understanding that humans and the environment are causally linked. Does this mean that some humans are intractably superstitious? Or does it mean that these ancient belief systems represent a plausible way of viewing reality? There is as yet no scientific proof to confirm the belief implicit in the words of the Indonesian woman, but scientists now think that the universe is a unified whole made up of energy. This does not prove a causal relationship, but it is highly suggestive of the unity of reality.

How humans integrate themselves with nature is part of the current scientific debate about climate change. But for hundreds of millions of indigenous people around the world, as for the Indonesian woman, there is no debate. Humans, ancient indigenous peoples believed, are not only an integral part of the environment but also play a prominent role in it. For most of those schooled in the scientific paradigm, however, the debate about climate change can be intense. Some in this camp believe that technology, although it contributed to the environmental devastation of the planet, will also contribute to its salvation. In other words, they believe that, as technology advances, it will be able to resolve every environmental issue, including climate change. This book is one attempt to contribute to the beginnings of an alternative answer. It does so by suggesting a form of constitutionalism that includes within its framework both humans and other life forms.

In the aforementioned story of Son of Raven, Ḱʷaaʔuuc the Creator kept the light in a box called *Huupakʷan'um*. The *Huupakʷan'um* is a symbolic representation of a way of life, embodying the supreme constitution for all life forms. This explains why a Nuu-chah-nulth *tyee Ḥaw'iɬ* (head chief) would have a *Huupakʷan'um*. The metaphor of "light" is both easy to understand and a complete mystery. From this metaphor, upon which an entire way of life can be based, there arose beliefs and practices that I refer to as *Haḥuuɬism* and that can now be articulated as an emergent form of contemporary constitutionalism.

Although the contents of *Huupak^wan'um* must remain unutterable (except as illuminated by metaphor), there is a need for a way of life that derives directly from this light. The quest for this light, made allegorically manifest in the story of Son of Raven, provided the ancient Nuu-chah-nulth with an appropriate way to negotiate reality. They found that this light enabled and illuminated as many lifeways and points of view as there are life forms without their society, dissolving into innumerable fragments. Thus, the Salmon people, the Bear people, the Eagle people, the Wolf people, the Cedar people, the Nuu-chah-nulth peoples, the Salish peoples, the Haida peoples, the European peoples, the African peoples, and the Asian peoples all have their own ways of life, their own points of view, their own written or unwritten constitutions. Is it possible, then, to begin to develop protocols such that all life forms may begin to enjoy a degree of well-being? This book attempts to begin an answer.

1

Wikiiš čaʔmiiḥta
Things Are Out of Balance,
Things Are Not in Harmony

On an ABC television news channel in March 2007, there appeared a documentary by the news anchorman Terry Moran.[1] For this documentary, Moran travelled from the contiguous United States to the tiny northern Alaskan village of Shishmaref, which is situated beside the Chuckchi Sea and north of the Bering Strait, just thirty-two kilometres from the Arctic Circle. The occasion for this trip was to investigate reports that an Alaskan village was being threatened by the effects of global warming. One can imagine that the only cultural memory of these Inupiaq people, who have lived in this area for millennia, must be one of a world covered in snow and ice for most of the year. Now it seems that the balance of ice and snow is shifting so as to affect both weather patterns and shoreline patterns.

The documentary shows a local man outdoors cutting meat and filling an old oil drum with snow. The oil drum sits on a large fire that appears determined to burn in spite of a prevailing breeze. Moran asks: "What are you cooking?"

"I got some fish and I got some seal meat."

Nayokpuk, the cook, explains that something funny is happening to the weather. Even though the surrounding environment seems covered with the ice and cold of winter, this is a deception. This year, in January, the weather is so warm that it is like springtime.

John Sinnock, a teacher of carving and traditional crafts at the Shishmaref School,[2] explains: "The ocean ice has been getting a lot thinner. It isn't as thick as it used to be. And it goes away much faster now than it did in the past when we were kids. In front of my mother's house, we used to look out at the gentle slope. In my life time we've probably lost about 400 feet." This kind of climate change in Alaska, where the fierce storms off the sea now last much longer than they used to because temperature increases have been so great over the past fifty years, means that the village of Shishmaref is gradually being destroyed. It now appears inevitable that the community must move to a safer location if it is to survive. The documentary concludes with questions for the high school students:

"How many of you want to leave your village?" Not a single hand goes up.

"How many of you want to stay?" They all do, although one or two do not seem to be sure, so they give in to peer pressure and put up their hands.

"Why do you want to stay?" The students answer that the village is their home, that life in the village contains all of their memories, that the village is their life.

Out on the west coast of Vancouver Island, some 2,414 kilometres south of the Alaskan village, I recall my attachment to my own community of Ahousaht, which is, like Shishmaref, completely isolated. Urban people often find it difficult to understand why anyone would want to live in these kinds of places. Can they really be telling the truth when people say that they do not want to leave?

During my residential school experiences in Port Alberni, it was common for everyone to want desperately to get back home. What is it that created that longing for home, regardless of its location? What is meant today when people speak of the "comforts of home"? For Aboriginal children in residential schools, the comforts of home did not refer to technological progress, the convenience of running water, indoor plumbing, dishwashers, washing machines, televisions, radios, computers, and iPods. No, the comforts of home for Aboriginal children have, until recent history, always been associated with the pre-eminence of relationships within the context and dynamics of place.

For example, from the perspective of my Nuu-chah-nulth heritage, living in community is not taken for granted because reality's inherent potential for creativity and destruction is accepted as natural.[3] The tension between creation and destruction is taken as *qua,* that which is, a given condition or state that translates into creative or destructive lifeways. Although life can at times be peaceful and pleasant, it can also be brutal, violent, conflicted, and destructive. It is for this reason that the sacred *ƛuukʷaana*[4] (remembrance ceremony) was created. The Nuu-chah-nulth believed that, in order to maintain community, in order to maintain a semblance of balance and harmony between beings,[5] people needed to be periodically reminded to be vigilant. The *ƛuukʷaana* can be considered a preventive measure performed by the community to appease the corrosive forces of reality. This preventive measure led to the common practice of *t'apswiis*, which is a form of early training for the *?uusumč*. This involved little children of around four years old running down to the beach before breakfast to dive into the ocean. *T'apswiis* was the first stage of training for the more rigorous *?uusumč*.

Diving into the cold waters of the ocean was a pleasant experience because it was always accompanied by the approval of the entire extended family. It also represented the kindergarten version of *haaḥuupa* (teachings) that were similar to contemporary mores and laws. Even during this earliest training and learning, Nuu-chah-nulth children participated in a community that went beyond the walls of their respective houses. Children ran naked down to the beach and touched the earth, clothed only by air, sometimes covered in early morning sunlight, and then were completely immersed in water.

The challenge of living in community also meant – unlike today's constitutionalism, which concerns itself primarily with human rights – the struggle to live in balance and harmony with all life forms, including a living earth that was known as Ḥaw'iḷume. The final syllable of this word, *ume* (pronounced oom-ee), means "mother." The word *Ḥaw'iḷ* has always been translated as "chief," but it also means wealthy. According to this usage, wealth is always inextricably bound to spiritual power. Consequently, a wealthy *quu?as* is any person, any life form, any being who has access to spiritual power. Ḥaw'iḷume, or Wealthy Mother Earth, is a *quu?as* of great spiritual power who, at our current point in history, is under great duress. For the people of Shishmaref, this duress means a gradual destruction of their homeland.

Like the Nuu-chah-nulth, the people of Shishmaref have a legacy of *himwiča* (or storytelling). Their stories are about their land, which is full of their own histories, about the great exploits of their grandparents and the great discoveries of their first peoples. It seems that all indigenous peoples have songs, dances, and ceremonies that accompany their stories. And always they have something that Nuu-chah-nulth call *tupati*,[6] or the empirical demonstration of spiritual power. It is this attachment to the land, Ḥaw'iɬume, represented by stories, songs, dances, and ceremonies, together with a contemporary history, that is bound up with the deep and powerful sense of home felt by the Shishmaref students.

A news documentary cannot convey this kind of importance-of-place in a segment of five to ten minutes. From the perspective of the Alaskan village people, how can high school children articulate the rich social, political, economic, and spiritual heritage that is at once personal in terms of identity, communal in terms of shared historical experiences, and concrete in terms of a land base that has, until recently, provided them with a secure living? The answer is that they cannot; however, in addition, it must not be assumed that their attachment to an isolated place also means that they are unaware of trends in the larger world. Can children, regardless of whether they are urban or rural, be aware of conditions external to their social environment?

Yes, educational research conducted across a wide spectrum of urban/rural schools in the British Columbia school system found that even children in Grade 4 can be aware of external societal trends.[7] And I have little doubt that, if the ABC news anchorman had asked these Alaskan high school students whether they understood why their village was being destroyed, they would all have shown themselves to be conscious of climate change.

The Shishmaref experience provides one example of an issue that has become a global concern. Unless the current progression of climate change is altered, the island nation of the Maldives, to take another example, may eventually disappear under a rising ocean:

To visit the Maldives is to witness the slow death of a nation. For as well as being blessed with sun-kissed paradise islands and pale, white sands, this tourist haven is cursed with mounting evidence of an environmental catastrophe. To the naked eye, the signs of climate change are almost imperceptible, but government scientists fear the sea level is rising up to 0.9 cm a year. Since 80% of its 1,200 islands are no more than 1m above sea level, within 100 years the Maldives could become uninhabitable.[8]

Of course, one of the difficulties of the issue of climate change is inherent to the very nature of climate. There have been, over large geologic time scales, natural climate changes during which ice ages have alternated with warming ages. Even within the cultural memory of the Nuu-chah-nulth peoples, there is the knowledge that west coast weather went through a very cold period some two hundred years ago and a balmy period some one hundred years ago. However, the scale of these more recent fluctuations did not threaten life on the planet in the same way that Shishmaref and the Maldives are threatened today.

Another difficulty to add to the dialogue about the threat of climate change to life on earth is made worse by members of a community of scholars who are willing to compromise sound research principles in exchange for money. The Canadian Association of University Teachers, a national voice for university teachers, recently promoted a book about psychopharmacology that illustrates this compromise of research integrity. Professor of Medicine David Healy writes:

In any area of science, dominant paradigms exert an influence. However, as psychoanalysis demonstrated, when a science has a commercial basis, those who make a living out of one point of view seem less likely to tolerate dissent than is normal in the rest of science. In psychopharmacology the trinkets and junkets of influence are an obvious part of the culture ... Responsible adults recognize that even university departments are businesses these days and that practitioners are heavily influenced by the government and third-party payers in addition to pharmaceutical companies.[9]

According to a group in the United Kingdom called Scientists for Global Responsibility (SGR), the primary research of corporate scientists has a *commercial basis.*[10] In a speech given to SGR on 13 March 2004, Dr. Stuart Parkinson outlined in some detail the increasing influence of corporations on science and technology. For example, in the United Kingdom the amount of research dollars that corporations now spend on their own scientists is more than twice what universities spend. Not only do corporations have their own scientists on their payrolls, but they also fund research and development at universities. One of the unfortunate consequences of this is that scientific and ethical principles may often be compromised. The "trinkets and junkets of influence" become "an obvious part of the culture," and scientific research that strives to reveal "truths" is negatively affected. The public is confused between those scientists who seek to satisfy their corporate sponsors and those who sincerely seek to investigate climate change. For example, Al Gore points out that, in 2006, the Union of Concerned Scientists, which is based in the United States, reported that "ExxonMobil [had] funnelled nearly $16 million between 1998 and 2005 to a network of 43 advocacy organizations that seek to confuse the public on global warming science."[11]

In opposition to those influenced by corporate interests are those who have become part of a global movement that Paul Hawken – environmentalist, author, and journalist – refers to as providing "blessed unrest."[12] Since it has no overarching leader or structure, it is a movement that is immune from corporate influence. Hawken explains:

These ... individuals ... are part of a coalescence comprising hundreds of thousands of organizations. It claims no special powers and arises in small discrete ways, like blades of grass after a rain. The movement grows and spreads in every city and country ... The movement can't be divided because it is so atomized ... It forms, dissipates, and then regathers quickly, without central leadership, command, or control ... It has been capable of bringing down governments, companies, and leaders through witnessing, informing, and massing ... Picture the collective presence of all human beings as an organism. Pervading that organism are intelligent activities, humanity's immune response to resist and heal the effects of political corruption, economic disease, and ecological degradation, whether they are the result of free-market, religious, or political ideologies ... The movement has three basic roots: environmental activism, social justice initiatives, and indigenous cultures' resistance to globalization, all of which have become intertwined.[13]

Although the phenomenon of "blessed unrest" is primarily a grassroots movement of people from every culture and continent, it may also represent a natural self-organizing principle free of any coercive hegemonic ideology. The philosophy suggested by "blessed unrest" is that creation is not governed by random selection but by something else. David C. Korten, an economist with a background in psychology who works out of Stanford University, explained in an interview with the *Sun* (magazine) that he came to a new understanding of living systems that presents a view of reality that differs from that proposed by Charles Darwin:

> When I was first looking for a model for a new economics, I looked to biological systems for the needed organizing principles. Our conventional understanding of living systems is the Darwinian theory of ruthless competition. Modern biologists, however – particularly female biologists such as Janine Benyus, Mae-Wan Ho, Lynn Margulis, and Elisabet Sahtouris – have discovered that living systems are fundamentally cooperative. Obviously there are competitive dimensions; Darwin didn't make that part up. But life can exist only in cooperative, sharing relationships with other life. Energy is constantly flowing back and forth among organisms, just as it is among the cells of a single organism. [14]

Self-organization appears to be a universal principle applicable to all levels and dimensions of reality, from the molecular to the cosmic. Each cell in the human body works on the principle of self-organization without taking coercive direction from any specific organ, yet the well-being and health of the whole is the apparent goal. This is a scientific interpretation of a healthy system such as that found in a well-balanced and harmonious human body. When Korten says that "life can exist only in cooperative, sharing relationships with other life," he is describing a healthy, sustainable system. For humans, it has been, to date, an insurmountable challenge to integrate the *competitive dimensions* of reality with the *cooperative dimensions* – a process that appears necessary for sustained life.

In addition to this global environmental movement, there are thoughtful observers, scholars, and scientists who are concerned about the underlying causes of our planetary crisis. For example, author and social critic James Howard Kunstler identifies economic theory and its practice as one source that contributes to the problem:

> The free-market part of the equation referred to the putative benefit of unrestrained economic competition between individuals, and because corporations enjoyed the legal status of persons, they were assumed to be on an equal footing with other persons in a given locality. Thus, Wal-Mart was considered the theoretical equal of Bob the appliance store owner, and if Bob happened to lose in the retail competition because he couldn't order 50,000 coffee-makers at a crack from a factory 12,000 miles away in Hangzhou, and receive a deep discount for being such an important customer, well, it wasn't as though he hadn't been given the chance.[15]

"Economic competition" is the business equivalent of Darwin's notion of "survival of the fittest." The economic inequity between Wal-Mart's owners and Bob appears to be sanctioned by a prevailing scientific story. In keeping with this scientific view of reality, Kunstler relates an insider story about the attitude of corporate executives towards their role in life:

> Colin Campbell, an oil geologist who has worked for many of the leading international oil companies, including BP (British Petroleum), put it this way: ... "It's not their job to look after the future of the world. Their directors are in the business to make money, for themselves primarily and for their shareholders when they can ... It's over! It's finished! And how can BP or Shell and the great European companies stand up and say, well, sorry, the North Sea is over? It's a kind of shock they don't wish to make. It's not evil, or there's no great conspiracy, or anything. It's just

practical daily management. We live in a world of imagery and public relations and they do it fairly well, I'd say." [16]

It appears that the theory of evolution and its notion of *survival of the fittest* may be reflected in the behaviour of some corporations. *It's not their job to look after the world.* In turn, the overwhelmingly competitive edge enjoyed by powerful corporations over individually owned businesses continues to result in devastation to the environment. In addition to these justifications for environmental destruction, there is also the fact of its amoral nature. David Korten, who spent five years on the faculty of the Harvard Business School, writes in his book *The Great Turning:*

> Professors of law and businesses commonly teach their students that bringing ethical considerations into corporate decision making is unethical, as it may compromise the bottom line and unjustly deprive shareholders of their rightful return. It is a rather perverse moral logic ... Consequently, under current US law, the publicly traded limited-liability corporation is prohibited from exercising the ethical sensibility and moral responsibility normally expected of a natural-born, emotionally mature human adult. If it were a real person rather than an artificial legal construction, we would diagnose it as sociopathic. [17]

As Korten's book is entitled *The Great Turning,* it should be noted that there is some hope that what is taught in business schools will change from an insistence that ethics in corporate decision making is unethical to an insistence that ethics performs a necessary function in the business world.

Although some aspects of the issues around climate change are subject to debate, others are already historical fact. In addition to scientific predictions about the limits of the oil supply, [18] which is fundamental to the maintenance of a technological civilization, there is, within local memories, widespread knowledge about the impact of climate change.

On the East Coast of Canada the cod fishery has collapsed, while on the West Coast the same can be said for the wild-stock salmon fishery. When I was a little boy, the inner harbour of my village seemed crowded with at least thirty boats that were licensed to fish continuously for six months of the year. As of 2009, that number is reduced to four or five that are licensed to fish for one or two days of the year. Roy Haiyupis, a distant uncle of mine, now deceased, served on the Clayoquot Sound Scientific Panel from 1993 to 1995. He recalled a certain bay in our traditional territory that was so full of sockeye that the collective sound they made when they finned and broke the surface of the water was like the sharp impact of a rifle. (Here, Roy would clap his hands together and, at the same time, say, "BOOM!") The rivers, bays, and estuaries that once produced salmon abundantly are now silent.

The United Nations Earth Summit, held in Rio de Janeiro in June 1992, articulated the Framework Convention on Climate Change, which was then followed by other agreements, such as the Convention on Biological Diversity and Agenda 21. Each of these conventions is explained in the report prepared by the Scientific Panel.[19] The panel was one initiative by one province in one country (out of at least 160 countries that were concerned with the earth's environment). It arose as a result of concern about the depletion of coastal temperate rain forests, which comprise only 0.02 percent of the earth's land mass. Fully one-quarter of the earth's rain forests are found in British Columbia, including Clayoquot Sound, which is within my home territory. Besides a diverse range of scientists, the panel included three Nuu-chah-nulth elders and one Nuu-chah-nulth academic who, together, could represent traditional ecological knowledge. Like the people of Shishmaref, who possess lived-experience of climate change and its impacts on their village, the three Nuu-chah-nulth elders were, with their collective lived-experiences, able to add to the growing global knowledge base about the devastation that human activity is causing to the environment.

As a co-chair of the Scientific Panel, I was able to see and witness, in the micro-world of BC environmental politics, a reflection of the conflict in the macro-world of the global scientific community. In one confidential meeting with high government officials, the co-chairs were explicitly requested to rubber-stamp extant government forest-practice policies. We refused. We said to them that, as a government, they should take the "heat" in the short term because it was certain that any panel recommendations would severely limit clear-cut forestry practices, which might not be popular. But they were assured that, in the long term, they would receive recognition for doing the right thing. How does this story reflect the current scientific debate about global warming?

First of all, the BC government's forestry policy and practice is based on stumpage fees. Since most of the forest area is on Crown land, the government is able to collect revenue from forest companies through a tree-farm licence system. The more trees that are cut, the more revenue there is for the government. Prosperity depends on clear-cutting the forests. As stated in earlier testimony by Colin Campbell, the only business of corporations is to make money for its shareholders and for its executives. It appears that what people and corporations want is in conflict with the environment's need for sustainability.

How Much Can Humans Know about Reality?

In terms of lifespan, humans have a recorded history of some five thousand years, but the physicist Frank K. Tippler estimates that, in terms of geologic ages, the universe is 20 billion years old and will likely exist for more than another 100 billion years.[20] Since physicists find no distinction between the nature of past, present, and future, our state of knowledge about the whole of this reality must necessarily be comparatively young and incomplete. What may eventually become clear about our "state of knowledge" is that humans, in our current stage, can know very little in comparison to what is not known. With respect to contemporary scientific knowledge about the climate of the whole earth system, James Lovelace, author of more than two hundred scientific papers, declares that he finds it extraordinary, "given the depth of our ignorance," that "scientists are willing to put their names to predictions of climates up to fifty years from now and let them become the basis for policy."[21] Here, Lovelace is referring to the work of more than one thousand scientists, from many different nations, who make up the United Nations Intergovernmental Panel on Climate Change (IPCC). One of the predictions the IPCC scientists made about the melting of Arctic ice does not match current rates of melt, which indicate that the Arctic may be ice-free during the summer within fifteen years rather than within fifty years.[22] Of continuing concern is the fact that some scientific predictions are routinely replaced by new findings and so their reliability is called into question. In other words, some of the scientific knowledge that humans claim to possess is inconsistent with reality. Moreover, non-scientists are expected to accept on faith much of what they are told by the scientific community. In the area of chemistry, biophysics, and other advanced scientific disciplines, as well as in the area of philosophical study, we, the general populace, live much as did illiterate people during the Dark Ages of Europe: we depend on the few to interpret for us the nature of "truths" and "realities." Although the gap between the knowledge held by the general populace and the specialized knowledge held by today's scholars remains large, this gap has been considerably narrowed through the common global experience of climate change. While complex and nuanced discussions by scientists and environmental philosophers are relevant and important, so, too, is this global experience of climate change. In other words, my proposition about human knowledge in the area of climate change is that both scholarly discourse and the global, experientially based knowledge movement (which is the direct result of climate change) are of equal importance. The global nature of climate change means that the subject cannot be the exclusive domain of scholars but, rather, through direct experience, includes a great variety of interest groups from every nation and culture.

While climate change remains a question among the scientific and scholarly community, it is not a question for those with direct (and threatening) experience of it. For the latter, the evidence for climate change is clear in its destruction of homelands, in its degradation of rivers that formerly spawned millions of salmon, and in its destruction of habitat with the consequent disappearance of many species. For the scientific and scholarly community, questions of climate change are purely theoretical, lacking any hope of concrete certainty. An immediate concern for this community is its credibility, which, unless it can demonstrate its pronouncements with practical outcomes, may eventually come into serious question.

The Prevailing Story that Guides the Current World Order

In *The Clash of Civilizations and the Remaking of World Order,* Samuel Huntington, chair of the Harvard Academy for International and Area

Studies, writes: "World views and causal theories are indispensable guides to international politics."[23] For most of human existence, the "world order" has been characterized by little contact between peoples and cultures separated by a geography made vast by the primitive level of technology. Then, in 1500, beginning with the voyages of Christopher Columbus and Amerigo Vespucci, which led to global colonization, the world order became Western. Eventually, during the twentieth century, the West became bipolar: the Capitalist West versus the Communist East. Today, in the post-Cold War era, the world order has come to be defined by cultures or civilizations. Among all the stories represented by different cultures and civilizations, there is one that predominates. That is the story of the West, represented principally by the United States.

In her examination of the story of the West, Jean Shinoda Bolen, a clinical professor of psychiatry at the University of California and a distinguished life fellow of the American Psychiatric Association, writes: "The mythology of a culture, in this case Western civilization, instructs us about the values, patterns, and assumptions on which this culture is based. When we stop to examine our mythological heritage, we may be enlightened or appalled by how much it is a metaphor for what exists in contemporary reality, how much our mythology is about us."[24]

In *Tsawalk: A Nuu-chah-nulth Worldview* I tell the story that guided my people until the Europeans arrived. The Europeans came with many stories of their own, but only one story provided an "indispensable guide" to understanding the current world order. This current world order, inasmuch as it has been created by the West, is a reflection of the "values, patterns, and assumptions" upon which it is based.

"It's All a Question of Story"[25]

So says Thomas Berry, adding:

We are in trouble just now because we do not have a good story … The Darwinian principle of natural selection involves no psychic or conscious purpose, but is instead a struggle for earthly survival that gives to the world its variety of form and function. Because this story presents the universe as a random sequence of physical and biological interactions with no inherent meaning, the society supported by this vision has no adequate way of identifying any spiritual or moral values.[26]

It *is* all a question of story. Although there are several versions of the evolution narrative, Berry presents the predominant one, which continues to be taught in all mainstream curricula, from elementary schools through to postgraduate programs. Berry thinks that most of the earth's current problems have been influenced by this story, which continues to predominate. It is a story that, over the past five hundred years, has reflected the principles of natural selection and survival of the fittest, according to which the powerful prevail and the weaker are either destroyed or dominated. The natural outcome – politically, socially, and economically – is an unequal distribution of wealth and political goods, a process of imbalances and disharmonies that guarantees conflict.

Drawing on the most advanced findings of physics, Fritjof Capra, physicist and systems theorist, has this to say about why and how we have ended up with our current global problems:

My starting point for this exploration was the assertion that the major problems of our time – the threat of nuclear war, the devastation of our natural environment, our inability to deal with poverty and starvation around the world, to name just the most urgent ones – are all different facets of one single crisis, which is essentially a crisis of perception. It derives from the fact that most of us ... subscribe to the concepts and values of an outdated worldview, to a paradigm that is inadequate for dealing with the problems of our overpopulated, globally interconnected world.[27]

Moreover, Capra points out that the current version of this scientific story

consists of a number of ideas and values, among them the view of the universe as a mechanical system composed of elementary building blocks, the view of the human body as a machine, the view of life as a competitive struggle for existence, the belief in unlimited material

progress to be achieved through economic and technological growth, and
– last, but not least – the belief that a society in which the female is
everywhere subsumed under the male is one that is "natural." During
recent decades all of these assumptions have been found to be severely
limited and in need of radical revision.[28]

In other words, the main story no longer appears to support certain assumptions about the nature of reality and is thus "severely limited," resulting in
a "crisis of perception."

The nineteenth-century philosopher Friedrich Nietzsche appears to have
understood this "crisis of perception" when, in 1882, he wrote:

Have you not heard of that madman who lit a lantern in the bright morning hours, ran to the market-place, and cried incessantly: "I am looking
for God! I am looking for God!"

As many of those who did not believe in God were standing together
there, he excited considerable laughter. Have you lost him, then? said
one. Did he lose his way like a child? said another. Or is he hiding? Is he
afraid of us? Has he gone on a voyage? or emigrated? Thus they shouted
and laughed. The madman sprang into their midst and pierced them with
his glances. "Where has God gone?" he cried. "I shall tell you. We have
killed him – you and I. We are his murderers. But how have we done
this? How were we able to drink up the sea? Who gave us the sponge to
wipe away the entire horizon? What did we do when we unchained the
earth from its sun? Whither is it moving now? Whither are we moving
now? Away from all suns? Are we not perpetually falling? Backward,
sideward, forward, in all directions? Is there any up or down left? Are we
not straying as through an infinite nothing?"[29]

On 8 April 1966, *Time Magazine*'s front page carried the question: "Is God
Dead?" The issue of God's existence remains as controversial now as it was
in Nietzsche's day.[30] Nietzsche's narrative is exquisitely ironic: "How were
we able to drink up the sea? Who gave us the sponge to wipe away the entire
horizon? What did we do when we unchained the earth from its sun?"
Metaphorically, these abilities are God-like because they imply the capacity

to create or reorder the nature of the physical universe and, by so doing, to impose purpose and design on creation. For Nietzsche, then, it makes sense – in the context of this new order of creation as interpreted by science – to ask: "Whither is it moving now? Whither are we moving now? Away from all suns? Are we not perpetually falling? Backward, sideward, forward, in all directions? Is there any up or down left? Are we not straying as through an infinite nothing?"[31]

According to David B. Allison, Nietzsche declares that God is dead because: "His function as creator ... was replaced by another agency, namely, by *science,* and by another faith – the faith and belief in an omnipotent *technology.*"[32] Erich Heller, in *The Importance of Nietzsche: Ten Essays,* writes: "The death of God he calls the greatest event in modern history and the cause of extreme danger. Note well the paradox contained in these words. He never said that there was no God, but that the eternal had been vanquished by Time and that the Immortal suffered death at the hands of mortals: God is dead."[33] Walter Kaufmann, a German-American philosopher, maintains that, on the one hand, Nietzsche is misunderstood in the English-speaking world and, on the other, that Nietzsche can be contradictory. According to Nietzsche:

> That we find no God – either in history or in nature or behind nature – *is* not what differentiates us, but that we experience what has been revered as God, not as "godlike" but as miserable, as absurd, as harmful, not merely as an error but as a *crime against life.* We deny God as God. If one were to *prove* this God of the Christians to us, we should be even less able to believe in him.[34]

Although Nietzsche is famous for saying "God is dead," the irony inherent in this statement seems to indicate that, in denying the existence of God, humans have cut themselves off from an important source of knowledge. This differs from the position of the early twentieth-century philosopher Clement C.J. Webb, who holds that civilization could not have begun until humans stopped believing in spiritual stories.[35] According to him, God and a belief in a spiritual reality are a barrier to knowledge.

Nietzsche's question is about who we are, about the nature of human existence, in the same way as is the scientific storyline. Thus, the interpretation of human nature can be approached from the perspective of the scientific story of evolution or from the perspective of the stories that preceded it. Riane Eisler, a cultural historian and evolutionary theorist, asks the following questions:

> Why do we hunt and persecute each other? Why is our world so full of man's infamous inhumanity to man – and to woman? How can human beings be so brutal to their own kind? What is it that chronically tilts us toward cruelty rather than kindness, toward war rather than peace, toward destruction rather than actualization?
>
> ...
>
> Yet, if we look at ourselves – as we are forced to by television or the grim daily ritual of the newspaper at breakfast – we see how capitalist, socialist, and communist nations alike are enmeshed in the arms race and all the other irrationalities that threaten both us and our environment. And, if we look at our past – at the routine massacres by Huns, Romans, Vikings, and the Assyrians or the cruel slaughters of the Christian Crusades and the Inquisition – we see there was even more violence and injustice in the smaller, prescientific, preindustrial societies that came before us ... Is a shift from a system leading to chronic wars, social injustice, and ecological imbalance to one of peace, social justice, and ecological balance a realistic possibility?[36]

These are valid and timely queries. Why *is* this inhumanity an apparently "chronic" condition that seems to thrive today, regardless of ideological persuasion, the same way as it did in earlier times? Is change possible? Eisler is an evolutionist as well as a feminist and social scientist, and she believes that the current "male dominator" model can be changed. In 2007, she published *The Real Wealth of Nations,* in which she proposes one avenue of change: "Failing to include caring and caregiving in the economic models is totally inappropriate for the postindustrial economy, where the most important capital is what economists like to call *human capital:*

people."[37] Eisler's dissatisfaction with, and concern for, the human condition is echoed in the writings of Patricia Marchak, a sociologist, and Jerry Franklin, an ecosystem analyst, who observe:

> We inherited a model of society in which all is subordinate to economic growth ... The economic model that has guided western [sic] society for several centuries offers no lasting basis for sustainability.
>
> ...
>
> Perhaps the greatest challenges of all lie in the realm of the social contract. Old understandings seem to be null and void: who, today, would champion the traditional role of a professional "priesthood" entrusted with resource management? Who would turn to science and technology as the ultimate sources of wisdom?[38]

If the "model of society" is one in which all is subordinated to "economic growth," it is not surprising that this poses a great problem for the social contract. The economic model, based on free enterprise and, by extension, on natural selection and survival of the fittest, seems to have created not only an environmental crisis but also a social crisis, in which the wealthy become wealthier and the poor become poorer. The economic model does not fit the social contract and, in fact, seems to weaken and even destroy it. Since science and technology, in their very brief history, have contributed to the current environmental crisis and lack of renewable energy, and since no solutions to these problems have yet been presented, one must indeed ask: "Who would turn to science and technology as the ultimate sources of wisdom?"[39]

Not the Story but Its Interpretation

In defence of the scientific enterprise, Richard A. Posner, a graduate of both Yale and Harvard, a prolific author, and now a senior lecturer at the University of Chicago, maintains:

> But unlike science, metaphysics lacks agreed-upon criteria for the evaluation of its theories. As a result, in an open, diverse, competitive culture,

the kind a pragmatist, being a Darwinian, tends to prefer, metaphysical disputation is interminable. This does not mean that the pragmatist "rejects" metaphysics. He rejects the possibility of establishing the truth of metaphysical propositions a priori; and it is in the nature of metaphysics that its propositions cannot be established empirically.

...

If, as his [Darwin's] theory implied, man had evolved from some ape-like creature by a process of natural selection oriented toward improved adaptation to the challenging environment of earliest man, human intelligence was presumably adapted to coping with the environment rather than to achieving metaphysical insights that could have no adaptive value in the ancestral environment.[40]

Posner's view, then, is that metaphysics is beyond any hope of systematic investigation because its propositions cannot be established empirically and do not have any adaptive value when placed within a Darwinian evolutionary context. From this point of view, God's death is irrelevant because science has no means of reliably investigating the issue.

The English environmentalist and author Jonathan Porritt thinks that the problem is not the scientific story but, rather, people's interpretation of that story:

Populist interpretations of evolution, from Hebert Spencer and Thomas Huxley onwards, have accustomed people to the idea of nature being "red in tooth and claw," with all life forms engaged in endless titanic struggles to ensure "the survival of the fittest." So what could be more "natural" than the history of humankind (both before and after the Industrial Revolution) being cast in the same metaphorical framework? This rationale of social Darwinism has been taken up with unbounded enthusiasm by the politicians and academic economists most centrally involved in the new-liberal revolution of the last 25 years.[41]

When considered as a classical scientific paradigm, the theory of evolution is an accurate description of the workings of nature. Nature has every outward appearance of being "red in tooth and claw," ensuring the survival of the most savage and barbaric. The wolf and lion do take down and devour their quarry; the eagle does swoop silently and swiftly to kill its unsuspecting prey; the strongest animals do get to mate so that their genes can be passed to the next generation. These are all empirically observable and scientifically testable, and they form an important part of human knowledge. However, this knowledge base, as implied by Thomas Berry, also assumes that these animals exist without conscious purpose, that their actions are performed within a context of non-communication and non-relationships.[42] It seems that Porritt's assertion that one of the causes of our global problems can be traced to "populist interpretations of evolution" must thus apply to the progress of human development and, by logical extension, to the social contract.

A Declaration of Triumph for Liberalism

Francis Fukuyama, professor of international political economy at Johns Hopkins University, presented the principles of liberty and equality that frame the social contract: "With the American and French revolutions, Hegel asserted that history comes to an end because the longing that had driven the historical process – the struggle for recognition – has now been satisfied in a society characterized by universal and reciprocal recognition. No other arrangement of human social institutions is better able to satisfy this longing, and hence no further progressive historical change is possible."[43] In other words, history comes to an end when the struggle between liberalism and other "isms" for political pre-eminence has been settled. To substantiate this claim, Fukuyama writes that, of "the different types of regimes that have emerged in the course of human history, from monarchies and aristocracies, to religious theocracies, to the fascist and communist dictatorships of this century, the only form of government that has survived intact to the end of the twentieth century has been liberal democracy."[44]

If Hegel's notion is taken to refer to universal history, and if this notion of history is placed within an evolutionary context in which life is believed to be continuously evolving – from simple to complex, from barbaric to civilized – then the aftermath of the French and American revolutions should have led to a world in which it could be said, as does Fukuyama, that "no further progressive historical change is possible."[45] Humanity is evolving towards a general state of liberalism that will be expressed in many languages, in many cultures, and in many worldviews, all of which will provide a framework for the social contract, for an agreement on how people will live together.

However, according to Noam Chomsky, history seems not to have come to an end, for the relationship of the United States – the pre-eminent superpower on the planet – to the United Nations (and, indeed, to many individual nations), like its "war on terror," is neither liberal nor democratic. Chomsky makes the following observations:

> Among the most salient properties of failed states is that they do not protect their citizens from violence – and perhaps even destruction – or that decision makers regard such concerns as lower in priority than the short-term power and wealth of the state's dominant sectors. Another characteristic of failed states is that they are "outlaw states," whose leadership dismiss international law and treaties with contempt. Such instruments may be binding on others but not on the outlaw state.[46]

Could it be argued that the *idea* of liberalism has triumphed but not necessarily the *practice* of liberalism? Can it be said that liberal states are characterized by an inability to protect their "citizens from violence" and even "destruction" or that such concerns are "lower in priority than short-term power and wealth"? Isn't an "outlaw" state the very antithesis of liberalism? On the one hand, there is Fukuyama's triumphant declaration that history has come to an end, and on the other hand, there is the United States, the most powerful liberal state, which, according to the basic principles of liberalism, is a "failed state." At the same time as this failed state violates fundamental liberal principles, it also represents itself as the champion of democracy: "No one familiar with history should be surprised that the growing democratic deficit in the United States is accompanied by declarations of messianic missions to bring democracy to a suffering world. Declarations of noble intent by systems of power are rarely complete fabrication, and the same is true in this case."[47] If Fukuyama's declaration refers to the triumph of the principle of liberalism and Chomsky's analysis indicates failure in the practice of liberalism, then one may conclude (as do Posner, Eisler, and other scholars) that the ideas generated from the Age of Enlightenment (also known as the Age of Reason) are not incorrect but merely incomplete.

Samuel P. Huntington, an American political scientist now deceased, provides further insight into the notion of incompleteness from the perspective of US foreign policy: "Religiosity distinguishes America from most other Western societies. Americans are also overwhelmingly Christian, which distinguishes them from most non-Western peoples. Their religiosity leads Americans to see the world in terms of good and evil to a much greater extent than others do."[48] If one of the most scientifically oriented and advanced states is also overwhelmingly Christian, perhaps the phenomena of "failed states" can be understood in terms of this apparent contradiction. Because the United States is an overwhelmingly Christian state with a declared moral mission, it presents itself to the world as a pre-eminent democratic role model, despite being, at the same time, a scientifically oriented state. In other words, accepted scientific theory is the story of who and what we, as humans, are, in the same way that Enlightenment ideas of liberalism is the story of what we think we are or should be. However, these stories are not in agreement with practice, and this is reflected in the foreign policy of the United States, where "short-term power and wealth" are given priority over principles of equality and freedom.

A Constitutional Issue

Apologists for and defenders of democracy and liberalism often reference two (*liberté, égalité*) of the three (*liberté, égalité, fraternité*) words that arose from the French Revolution and formed the basis of the French Constitution. Given the current conflicted state of the globe, the cry of *fraternité* appears to be just another Utopian and Enlightenment form of romanticism. Anthony J. Hall, founding coordinator of globalization studies at the University of Lethbridge, finds the contradiction between what is implied by *liberté, égalité, fraternité* and the current state of US foreign policy to have disturbing implications. According to Hall: "[In] mounting its War on Terrorism in the wake of the September 11, 2001, tragedies, the regime of George W. Bush drew heavily on the evangelical impulse of the West's old civilizing mission ... to describe the real, illusory, or manufactured enemies of the American way of life."[49] Part of the problem is that the global conflicts involving the world's contemporary superpower appear remarkably similar to the conflicts that wrought so much destruction in the recent colonial era. To make this point, Hall draws on the specific wording of the US Declaration of Independence:

> The War on Terrorism has deep roots in American history that cut far beneath the events of September 11. While the labels of the demonized other may have changed over time, the imagined attributes of the stigmatized foes of the American Dream have remained remarkably consistent since the era of the founding of the United States. In their very first act of self-justification, the founders carved out in 1776 a special category to encompass a class of humanity deemed bereft of inalienable rights, a class of people thought to embody such potential for unpredictable violence and anarchy that they were placed outside the assertions of equal rights proclaimed as the raison d'être of the revolutionary republic ... The War on Terrorism gave renewed force and legitimacy to prejudices similar to those that once induced the authors of the Declaration of Independence to refer to the Indigenous peoples of North America as "merciless Indian savages" ... From the beginning of this ascent, those distinct peoples who stood in the way of the United States' territorial ambitions were dehumanized and criminalized in the text of the Declaration of Independence.[50]

Is it of any significance that, underlying the terminology of the twenty-first century, are meanings and constructs relating to the "other" that have their roots in the Enlightenment? On the one hand, from these roots come declarations of freedom; on the other hand, from these same roots come the dehumanization, criminalization, and enslavement of the peoples of an entire continent. This historical fact is so recent that its negative impact in the form of socio-economic dysfunctions within indigenous societies continues to the present day. The German philosopher Immanuel Kant asked rhetorically: "What is Enlightenment?" To which he replied: "Enlightenment is man's emergence from his self-imposed nonage."[51] Kant then makes clear what he means by "man": "And then will not the European population in these colonies, spreading rapidly over that enormous land, either civilize or peacefully remove the savage nations who still inhabit vast tracts of its land?"[52] Enlightenment ideas of freedom and equality were limited, which is one way of saying that they were incomplete.

It is difficult not to conclude that the manufacture of the current world order in terms of a "war on terror," in terms of "us" and "the other," derives directly from ideas developed during the Enlightenment. This war on terror appears to be the outcome of a colonial agenda, the point of which is to civilize the peoples of the world. Instead of freedom and equality there is global conflict.

A Human Development Metaphor

The psychologist Chellis Glendinning, who is also a pioneer eco-psychologist, questions the problems of humankind from a human development perspective: What is the presenting symptom that characterizes the current world order? Chellis declares that it is *separation and divisiveness*.[53]

The context of human health in which the question is asked can be described in terms of three dimensions of consciousness. The first is the I-in-We dimension, which provides a necessary sense of belonging, security, trust, and faith in the world. Without this, children have been known to die. During the thirteenth century, the Holy Roman Emperor, Frederick II, wanted to determine humankind's original language, and he attempted to do this by ensuring that the children he was raising would never hear human speech. All the children died.[54] They were not allowed to experience the I-in-We dimension of consciousness.

In their discussion about love, three long-time collaborating psychiatrists, Thomas Lewis, Fari Amini, and Richard Lannon, explain what happened to those children under the care of Frederick II:

> But because human physiology is (at least in part) an open-loop arrangement, an individual does not direct all of his own functions. A second person transmits regulatory information that can alter hormone levels,

cardiovascular function, sleep rhythms, immune function, and more – inside the body of the first. The reciprocal process occurs simultaneously: the first person regulates the physiology of the second, even as he himself is regulated. Neither is a functioning whole on his own; each has open loops that only somebody else can complete. Together they create a stable, properly balanced pair of organisms. And the two trade their complementary data through the open channel their limbic connection provides.[55]

It seems that an *I* cannot develop and grow outside the context of a *We*. The second dimension of consciousness becomes possible as a direct result of a nurturing I-in-We environment. It is characterized by a struggle for healthy growth that results in a sense of personal integrity, a sense of value and worth, and a sense of purpose. Finally, the third dimension is the capacity to draw vision and meaning from non-ordinary states of consciousness. In terms of these three interrelated dimensions of consciousness, which complement each other and together make a unified whole, the presenting pathological symptom of the current world order is separation and divisiveness. This symptom, according to Chellis, can be traced back to at least the 1500s, when religious persecutions intensified into witch hunts. Hundreds of thousands of people, mostly women, mothers, and healers, were hanged, drowned, or burned in the town squares all across Europe.

Mass murders indicate a detachment from God-consciousness (a version of Nietzsche's "God is dead"), a separation of the human psyche from the compassion of soul and spirit. This psychic detachment and isolationism travelled to the Americas, where, given the previous practice and training in mass murder via the witch hunts, it produced a slaughter of numerous life forms – indigenous peoples, plants and forests, wildlife, and creatures of the sea – so immense that the pain of Ḥaw'iłume, Wealthy Mother Earth, seems now no longer bearable.

If humankind is experiencing a crisis of perception, as Capra states, and a problem of recognition, as Fukuyama suggests, then another significant part of the problem may lie with the human capacity to attain a sense of superiority. George Soros, whose credentials in this area stem from his experience as a very successful businessman and stock investor, which made him one of the world's richest people, writes: "Mankind's power over nature has increased cumulatively while its ability to govern itself has not kept pace."[56] He argues, moreover, that "the Age of Reason ought to yield to the Age of Fallibility. That would be progress."[57]

Son of Raven would agree. It appears that technological progress has been equated to, and mistaken for, human development.

A Nuu-chah-nulth Perspective

With regard to the lived-experience of the Nuu-chah-nulth and our consequent perspective on global issues, it is an understatement to observe that the planet is confronted with a crisis of perception, a problem of recognition, and a question of story, or to observe that there may have to be some rethinking of the social contract. We are, after all, continuing to suffer the tragic consequences of contact. In 1860, the Englishman Gilbert Sproat promised Ḥaw'ił Shewish of the Tseshaht people that the English were coming to improve the Nuu-chah-nulth way of life.[58] But a better quality of life for the Nuu-chah-nulth has not materialized.

I think that my great-grandfather Keesta, if asked about global issues, would thoughtfully comment that *wikiiš ča?miiħta* (things are out of balance, things are not in harmony). This is a phrase that can refer to any part of creation, including people, forests, animals, and all other life forms – the river, estuary, inlet, or ocean. Within Nuu-chah-nulth culture, the idea of balance and harmony is derived from origin stories.

Each story typically illustrates an aspect or interpretation of the nature of reality. The story of Son of Raven indicates that all of reality, the physical and the non-physical, is a unity. In the language of quantum mechanics, *nothing is separate from anything else. In Tsawalk: A Nuu-chah-nulth Worldview,* I explain the trial-and-error process that Son of Raven and his community undergo in their struggle to secure the light. What becomes clear from this process is that they do not at first know that creation is a unity. Their first assumption, made clear by their methods, is to assume that the community of wolves that owns the light has no interest in sharing it. After several failures to secure the light, it is Wren, the one who always speaks rightly, who provides the solution. It is a solution that I refer to as the *insignificant-leaf approach* or the *humble-stance approach* towards the non-physical or spiritual realm.

Origin stories help us to recognize that the unity of creation is diverse rather than hegemonic. Raven, Deer, Wren, and the entire diverse community of beings cooperate and work together, recognize each other, and practise the democratic principles of mutual consent and consensus building without compromising personal integrity. They are able to do this in spite of the fact that, on the surface, reality appears to consist of fragmentation, opposition, contradiction, and numerous mysteries and ambiguities. It was on the basis of this surface appearance that Son of Raven and his community first approached the Wolf community. They knew about the spiritual realm, that it was the domain of the Wolf community, even though they could not physically see it. And it was natural for them to assume that what could not be seen was separate from what could be seen.

The Nuu-chah-nulth discovery of the unity of all creation determines the nature of questions about the state of reality. Whereas scientists must ask why,[59] Nuu-chah-nulth peoples, confronted with the unutterable wonder of creation, are compelled to ask how? How does anyone describe the indescribable? How does one live and negotiate this creation? Since there definitely appears to be a kind of order, a method to the madness, how do we fit into the scheme of things? What can be said about those things that are forbidden to human knowledge systems, except to say that there are some *?uusumč* (vision quest) experiences of the non-physical that are not to be revealed, explained, or articulated? But how does one navigate what it is permissible to know? How does one balance and harmonize the disparate and contradictory elements of reality? How does this wondrous creation work?

In theory, the knowledge acquisition process of the *?uusumč,* which allowed the Nuu-chah-nulth to see beyond the purely physical reality of nature, also allowed them to discover that creation is a unity in spite of its apparent fragmented appearance and contradictory nature. For example, although deer, wolf, and salmon are scientifically classified as animals within the biological dimension of existence and therefore as separate from humans, Nuu-chah-nulth peoples also know and experience these animals as *quu?as,* as people like themselves. The same is true of trees and the multitude of other life forms.

What this means is that Nuu-chah-nulth peoples had to find some way to live with these other *quu?as* who were recognized as life forms, as living beings who were originally part of one language and community. Until the arrival of Europeans on Nuu-chah-nulth shores, the task of achieving balance and harmony between various life forms – between wolf and deer, between Nuu-chah-nulth and salmon, and between Nuu-chah-nulth and Nuu-chah-nulth – had been hard won through the development of protocols. Living in balance and harmony with diverse life forms is one way of describing a mature ecosystem. This principle of balance and harmony is necessarily applied to every dimension of existence.

For example, the story of Son of Raven and his community can also be interpreted as the development of a protocol for dealing with the physical and the non-physical. It was learned that access to the storehouse of the non-physical realm can be achieved not via the egotistical approach but via the insignificant-leaf, or humble, approach. Insignificance here translates into both a moral dimension (defined as humility) and into a natural description of human identity in relation to an infinite universe. A natural identity, when realized and enacted, can also translate into a key capable of unlocking mysteries or capable, like a well-focused lens, of revealing that which is. Son of Raven indicated this natural identity when he became an insignificant leaf, which became the key to gaining access to the storehouse of the non-physical domain. This discovery of a natural identity enabled the process of growth and development towards a completeness-of-being known to psychology as maturation. Just as any mature ecosystem will attain balance and harmony between its different life forms, so, too, people may attain balance and harmony between the physical and non-physical domains. But only after a purposeful struggle and only after the key is found: without a key the alternative is imbalance and disharmony.

When an earthquake erupts it can create the suspicion of *wikiiš ča?miiḥta* (imbalance or disharmony). In other words, as a force indivisible from the rest of nature, an earthquake is a sign of imbalance and disharmony. It is for this reason that the tsunami of 26 December 2004 elicited from that Indonesian woman the cry: "What did we do wrong?" In this cry, she recognizes the integration of human behaviour with the behaviour of the earth and its environment. These kinds of beliefs were labelled superstitious by early scientists and, perhaps, are so labelled by most contemporary scientists, although things may be slowly changing in some parts of the scientific community. For example, in a study of the discovery of the universal laws of advanced physics, Brian Greene notes:

> Physicists describe these two properties of physical laws – that they do not depend on when or where you use them – as symmetries of nature. By this usage physicists mean that nature treats every moment in time and every location in space identically – symmetrically – by ensuring that the same fundamental laws are in operation. Much in the same manner that they affect art and music, such symmetries are deeply satisfying; they highlight an order and a coherence in the workings of nature.[60]

The context for these symmetries is, of course, the quantum field that is present everywhere in space. These observations about the nature of our universe do not confirm the truth of the words of the tsunami victim who asked: "What did we do wrong?" But they do provide a sound theoretical basis, from a Nuu-chah-nulth perspective, for the view that, since there are universal symmetries with respect to physical laws, there may also be universal symmetries between the physical and the non-physical realms. Given this symmetry between the different domains of reality, there may be a direct relationship between the behaviour of people and the behaviour of the earth. Experiences derived from the ?uusumč indicate that, since reality is a unity, what befalls one part of this unity must befall the whole.

The Nuu-chah-nulth worldview, in keeping with the utterance of the Indonesian tsunami victim, sees the cause of the global crisis in relational terms, in terms of a creation filled with mutually interdependent life forms that require mutually acceptable protocols in order to maintain balance and harmony. From this point of view, the global crisis is one of relational disharmony. One life form, the human, has critically disrupted the balance

between other life forms and systems: the animals, the air, the forests, and
the seas. Ḥaw'iⱡume, Wealthy Mother Earth, like any wounded life form,
now seems quite naturally to be in the throes of fighting back, of roaring in
severe pain and anger, in response to the behaviour of the human life form.

2
Mirrors and Patterns

The myths, therefore, were the pattern ...

– Carobeth Laird, Mirror and Pattern[1]

No one today seems to know the meaning of Keesta.[2] What elders know, however, is that Keesta is a big name that belonged to one who is ʔuuštaqyu. Consequently it is a fitting name only for those who have become, in the language of human development, a mature person. ʔuuštaqyu means "one who has become complete." Its literal translations are: "one who has been fixed," "one who has been worked on," and "one who has worked on self." In cultural context these translations take on meanings that are multidimensional. The latter phrase, "one who has worked on self," may be the easiest to understand.

The former phrases – "one who has been fixed" and "one who has been worked on" – may be subject to misunderstanding. In a Nuu-chah-nulth context, these phrases always take on a cooperative and democratic meaning rather than a sinister meaning. Since both creative and destructive powers are available to vision seekers, the completed person is necessarily associated with creative powers, which, in turn, are characterized by principles that develop and enhance healthy relationships between life forms. In Nuu-chah-nulth, ʔuuštaqyu is a dynamic word that evokes, within the listener, heroic struggles of facing up to fear, facing up to natural and real threats, and it includes a severe test of faith and endurance until communication is effected between the vision seeker and the powers of the nonphysical domain. "One who has been fixed" and "one who has been worked on" refer to a co-scripting of reality on the part of the vision seeker and the non-physical domain.

Consequently, one who seeks knowledge, gifts, and powers through the ritual of a *ʔuusumč* is "one who works on self." When there have been, after many years of persistence and endurance, positive responses to this seeking of knowledge, gifts, and powers, it can be said that "one has been worked on, one has been fixed," because there has been a transformative and collaborative fusion of powers. Balance and harmony between the non-physical and physical have been achieved. A gift, power, or significant information has been granted, and this means that, in that instance, creative forces prevail both in the world of spirit and in the world of flesh. It is both a rational and a numinous experience, and its greatest expressions are not achieved by many. With this understanding, I think that I have a sound theory about the meaning of Keesta, a great name related to an achievement of balance and harmony and thus also related to universal principles of self-organization.

It was not uncommon for people who performed great exploits to share great names. Kleesh-in, Ah-Up-Wa-Eek, Maquinna, Wickinninish, Utlyu, Cupcha, Queesto, and similar names can all be shared by people who have performed similar exploits because Nuu-chah-nulth names, like plant nomenclature, simply reflect observable characteristics demonstrated by those who have been successful in the *ʔuusumč*. Kleesh-in, from the Ahousaht line of *hawiih,* can be traced to the fifteenth century, and it is also a name shared with and by other Nuu-chah-nulth. Queesto is a name that goes back fifteen generations and that latterly belonged to a Pacheedaht chief who spoke the Dididaht dialect of the Wakashan language group. Queesto seems to have the same root as *kʷist,* which means to change, to make different, or to transform. It is reasonable to surmise that Keesta has the same root as *kʷist* because, from the very beginning, transformation was a major theme in the negotiation of reality. Quis-hai-cheelth, a name given to my son Ahinchat when he was a young man, means "one who can change or transform." Quis-hai-cheelth, too, has the same root as *kʷist.* Son of Raven, who provided the model for *ʔuusumč,* practised transformation as naturally and easily as people today change their shirts.

Since Keesta grew up in accordance with the ways of his ancestors, and since I was born into his house, I, too, grew up with these ways, except for some cosmetic modifications taken from the Western technology of the day that served to enhance rather than to contaminate Nuu-chah-nulth lifeways.[3] Consequently, even though I was born some 448 years after the landing of Columbus, I was born into my ancestral ways. Moreover, my birth coincided with the beginning of the Second World War, which suspended intervention into indigenous ways for most of the 1940s. As a baby and toddler, I was treated and trained in the same way that Keesta had been when he was a baby and toddler. Typically, songs of identity, encouragement, acceptance, and future possibilities would be sung to me as I slept or as I was held in the arms of parents or grandparents. Songs of orientation for babies were common to each household. The loving, caring, fondling, and touching were accompanied by an understanding of the systematic logic behind the development of an identity. The formation of what contemporary educators term "supportive scaffolding," or "aid," which normally begins when children are of school age in Western societies, began at birth in our house.

Flores Island is located twenty-four kilometres north of world-famous Long Beach, which lies between the little maritime hamlets of Tofino and Ucluelet. Today, this area is well known as a tourist destination that, each year, attracts a few million visitors from around the world. However, when I was born, it was an isolated area without public roads. Once a month or so, the *Princess Maquinna,* a passenger and freight boat designed and built in 1912 for the rugged West Coast, created little buzzes of local excitement when it docked at all the little places and habitations. Other than this infrequent incursion from the outside world, the village of Ahousaht, where I was born, remained much as it was before the arrival of Captain Cook.

A story from my early life is that my father spoke to me soon after he drowned. I was eighteen months old and have no personal memory of it. However, what is significant is that this kind of story was commonplace within Nuu-chah-nulth culture during my time. In our house the story was often told by Nan. She said that my father told me, in our Nuu-chah-nulth language, that "everything will be alright." A few years ago, before she died, I asked Aunt Nora about this story. She took on a serious demeanour as she affirmed the story. Sometime later I phoned her younger sister, Aunt

Trudy, and asked about it. Trudy also affirmed the story and said that my father had spoken to me not once but several times and that each time the message was the same.

After the death of my father, according to ancient custom, they burned our big house and all of his personal effects in order to facilitate his transition from this world to the next. Did that ancient practice actually happen in my lifetime, fully 163 years after Captain Cook's landing at Yuquot in what is now Nootka Sound? Why are people surprised at this? Perhaps it is helpful to view this through the lens of classical science. For example, in biology, Willis W. Harman and Elisabet Sahtouris point out that "equilibrium is expected; change is to be explained."[4] "Equilibrium" is equated with stability and when applied to living societies solidifies them into unrealistic freeze-frames. Although this scientific view of reality is now being altered to fit more closely with a dynamic, living universe-of-being, the old stereotypes are still strong. As recently as the late 1980s, I attended a showing of a film of Tla-o-qui-aht performing a gift-bearing song and dance.[5] Since the dancers wore contemporary clothing under their regalia, someone commented that "you can't call *that* traditional!"

If one is informed by an outmoded scientific paradigm, the wearing of non-traditional clothing creates an impression of change (as expressed in the above comment). However, if one is informed by one's ancestors from within a *living* culture, which has, over millennia, recorded through story that change is not only natural to creation but also necessary to life, then the wearing of "shirts and blue jeans" during a dance *is* traditional. In the case of my father, then, even though his family wore contemporary clothing, the burning of his personal effects and the burning of the big house were conducted according to ancient laws and customs following the death of an heir to a chief's seat. In addition to this view that surface, or physical, changes are natural to creation, there are other kinds of change that are also natural. As a result of their explorations with the *ʔuusumč,* Son of Raven and his community changed from believing that reality is fragmented and intentionally unintegrated to believing in the unity of creation.

My first clear memories are not of the big house in which I was born but of the smaller nuclear-type home that replaced it after my father died.[6] It is from this smaller home that I would *t'apswiis* – run down to the beachfront before breakfast and dive into the ocean. *T'apswiis* is the beginning of early childhood training in preparation for the *ʔuusumč*. The experience of *t'apswiis* was a most pleasant one. Bathing in cold water has been a common practice among Nuu-chah-nulth for millennia, and, very possibly, over this time, there may have been some genetic adaptation towards a high level of tolerance for cold. In any case, for a little boy, this experience is all the more pleasant because of the general approval it evokes from all family members.

As a small child I was unilingual but adults were not. Not only were they fluent in our own dialect and in other Nuu-chah-nulth dialects, but some were also fluent in the trade language known as Chinook. In addition, they all had, to varying degrees, a basic working knowledge of English. In other words, Nuu-chah-nulth peoples were multilingual. Even within a common Nuu-chah-nulth dialect there could be variations of expression that illustrated a preference for creativity over uniformity. For example, *kʷatyiik* means "heavy" in the Ahousaht dialect, while *yuxyiik* means "heavy" in the Muwachat dialect. Despite the identical meaning of these two words, they have different (albeit similar) cultural referents. The cultural referent for *kʷatyiik* is a powerful supernatural being known as *Ǩʷatyat*. The spiritual power of *Ǩʷatyat* is translated into physical heaviness and becomes *kʷatyiik*. The cultural referent for *yuxyiik* is geographical: it is the summer home of the Muwachat, Yukʷat, which means "place of wind" as it is exposed to ocean breezes during the summer and strong winds during the winter. Wind, in our culture, has spiritual overtones. Thus, the meanings of *kʷatyiik* and *yuxyiik* are both simple and complex.

According to my first memories of my community, those who spoke primarily English were the exception. By a large margin, the language of preference was the language of our ancestors. My grandmother, Nan, spoke what was called, at the time, broken English, although she later became fluent. It is with warm amusement that I recall her daughters, my aunts, insisting that when they went to Tofino to shop they would do all the talking in English. Not long into the purchasing transaction, my grandmother would leap into the conversation, addressing the store clerk with her determined but broken English. Later, to everyone's merriment, the daughters would regale the entire household with this story.

Nuu-chah-nulth is my first language, and, as a child, whenever I heard the strange sounds of English I invariably found them to be harsh and alienating. Similarly, I think English speakers might find that the Nuu-chah-nulth language sounds harsh. For me, the Nuu-chah-nulth language is comforting, like familiar food, and is a natural way to make meaning. I still recall, in Grade 10, when my English teacher became ill and was replaced by a social studies teacher. For the first time, the English language began to make some sense to me. The social studies teacher taught English in a simple way, as is done in elementary school, and all of a sudden some dim lights began to dawn for me. No, I was not in a special class at the time, but rather in the academic stream. How did I manage? I managed by rote learning. I memorized everything without necessarily understanding it. How can I explain that I would often hear an English phrase and it would register as a blank in my mind? Often, during my time at the Alberni Residential School, someone would say something to me in English and I would not understand what had been said. The words would be just a jumble of meaningless sounds. As one can easily imagine, my inability to understand what was being said was taken as a sign of stupidity. I was often "stupid" when it came to understanding the English language. Even later, while at university, I would sometimes have to read a passage over and over and over because I could not comprehend it.

In hindsight, I can see that I was fortunate to have been born into, and to have lived completely within, the ways of my ancestors – their language, teachings, stories, practices, and form of governance. My grandfather took me to his council meetings, and I would sit with him and observe the proceedings. He never told me that I should pay attention, and he never spoke to me explicitly about this ancient way of governance. He simply took me with him. This is one of our methods of training. We did not wait for advanced age to learn political science and how to analyze various methods of decision making. Leadership training began early in life. What I observed during my grandfather's meetings has been misleadingly termed "consensus decision making." My grandfather's council meetings might sometimes have reached a general agreement on an issue, but just as often there would be differences of opinion.

I will illustrate with a case. It was decided, during the 1950s, that something should be done for the children of the community and that this should involve building them a play space in the form of a platform made of local wood. One man disagreed. He thought that the natural environment offered sufficient play spaces for the children. The natural environment – the beaches; the forests; the grassy fields; the tidal pools; the rocky shorelines covered with seaweed, kelp, and barnacles; the bullhead fish in the shallow waters of the inner harbour; the perch under the wharves; the little play canoes and much, much more – had, for millennia, been more than sufficient as a play area for little Ahousaht children. However, the vast majority agreed that a wooden platform was a good idea. The very next day, some men went out for the logs and began to assemble this platform. Among those working to build it was the man who had disagreed, and he made it abundantly clear that he did not think it a good idea. Wasn't that strange behaviour, given that it is more common today for those who disagree with a government decision to attempt to discourage and discredit it?

In the end, when the platform was built and the children were seen to play happily on it, the sceptical man again openly declared that he had disagreed with the idea, but he then added that he now saw that he had been mistaken. This man was one of the leaders in the community, and he offered a fine example of leadership. In this society people were trained to be leaders and to understand that much of life remains a mystery, as is indicated by ʔuusumč experiences. Further to this, daily experience and observations clearly indicated the interdependent nature of reality. One consequence of this was that it was held that if one did not ask for help when help was required, then one was not friendly. Although it must be remembered that no society has ever perfectly fulfilled its teachings and visions, the Nuu-chah-nulth generally valued and respected all life forms.

Nuu-chah-nulth teachings translated into a most remarkable decision-making behaviour – one that required none of the modern demands for a majority vote based on logical argument. Instead, what took place can be broken down into three simple rules: (1) each person has the right to a say; (2) each person has the right to be heard, and; (3) each person has the right to be understood. In the context of a culture that valued respect, not in an

ideological but in a spiritual sense, it seemed easy for each person to practise these three simple rules. There was no debate, no argument, no inter-
ruptions, no point-counterpoint, but rather a process whereby each person
in council exercised these three simple rules. If you think about this process, you realize that it takes much patience, much self-control, and a great
deal of effort to listen to and to understand each speaker, especially as today repetition is thought to be a waste of time. In ancient Nuu-chah-nulth
decision making, repetition was thought to be necessary to a sense of unity
and a sense of harmony. The need for each speaker to be understood also
minimized the likelihood of misunderstanding, and when a decision was
made in this way it became as strong as the number of people that made it.
This process of decision making did not eliminate differences of opinion
but it did provide confidence in any decision made. Here the metaphor of a
rope is useful: one strand of rope is easily broken, two strands less so, and
three or more strands still less so. The strength of these kinds of decisions
was virtually unbreakable.

A critical part of this governance system is the *ƛuukʷaana* (we remember
reality) ceremony. During the mid-1940s, a *ƛuukʷaana* was held in my village, and I did not realize at the time that Canada had made this ceremony
constitutionally illegal in 1884. When I was a little boy, the *ƛuukʷaana*
ceremony was simply a necessary part of my existence. The big house in
which it was held sat next to our house. When I approached the entrance, I
was immediately escorted to a seat next to my grandfather. I discuss a more
contemporary *ƛuukʷaana* in *Tsawalk: A Nuu-chah-nulth Worldview*, but its
main principle can be described as a form of polarity management, or reality management. However reality is defined or experienced, who can disagree, in this day of environmental degradation, terrorism, overpopulation,
and mass starvation, that it requires some kind of management? For the
Nuu-chah-nulth, the public ceremonial form of such management was, and
to some degree continues to be, the *ƛuukʷaana*. During this ceremony,
Nuu-chah-nulth are reminded that the loss of children (the "dissolution of
family values or the quality of family relationships," in contemporary
terms) to hostile or destructive forces can be ascribed to a failure to observe
the well-known teachings of our ancestors, who had a working knowledge
of, and experience with, the difficulties of a polarized existence. From a

Nuu-chah-nulth perspective, the concept of family also applies to all life forms that have families, such as the salmon, the deer, the bear, and the eagle. Indeed, ultimately, all of reality is considered to be a universe of (family) relationships.

But what is the *ƛuukʷaana* like from the perspective of a small child? All the adults within the family take on an air of fearful solemnity, as though they have become aware of some inevitable, pending danger. The air becomes so charged with this fearful solemnity that a *ƛuukʷaana* may be described as a community-arresting ceremony. Once everyone is seated in order in the big house, the doors are shut and guarded by *wit wok,* a police force that operates much like a national security brigade. Wolves are heard outside, pounding on the walls and attempting to break down the door. Fear spreads throughout the darkened hall. Somehow, in some supernatural way, the wolves are successful, and children, who have been selected beforehand, are stolen and lost to family and community. It is from the story of Son of Raven that all Nuu-chah-nulth learn that the wolves populate the community of the Creator, who owns the light. One of the great mysteries for the Nuu-chah-nulth is the question: since the chief Wolf owns the light and is therefore the giver of life, how is it that wolves also function as destroyers of families?

For a little child, it does not matter in the slightest that the ceremonies appear to be "pretend." For example, one of the dances is a bear dance, but under the bear costume human feet are clearly discernible. Later, I would learn that the ceremonies are never "pretend" as they are always about the nature of reality, about the consequences of ignoring teachings. In this sense, the *ƛuukʷaana* is similar to one of the traditional ritual practices of the South African Khoisan, who, at night, face up to the realities of their polarized existence by burying themselves in the earth up to their necks. Predators approach the heads of the Khoisan, which are seemingly resting on the earth, sniff them, and then trot off into the night. It is as if each animal understands the meaning of the Khoisan ritual because it does appear to result in a kind of peaceful co-existence. Mutwa, an African shaman, declares from personal experience: "If we humans can overcome this thing called fear, we can overcome the ills of this world and live in harmony."[7] So long as the Khoisan overcome their fears, they are safe from predators;

and so long as the Nuu-chah-nulth overcome their fears and face up to the predation of the sacred wolves, they are able to restore the bonds of their community.

There seems to be a principle to the polarity of existence, within inter-related contexts, that enables us to resolve problems when we have some understanding that the dark is there to enhance the light, the obstacle is there to strengthen the spirit, the resistance is there to enhance the freedom, and the fear, when confronted, is there to bring harmony. This principle does not apply when the polarity is overwhelmingly out of balance. I once worked with an architect who had emigrated from Czechoslovakia to Canada. While we worked on my reserve, he was invited to dinner by a lo-cal resident. He appeared uneasy, maybe even frightened, and asked me what he should do. "Go!" I said. He went and had a great time. He em-braced his fear, he buried himself up to his neck, so to speak, and experi-enced harmony in the creation of new friends. This story is an overly simplified explanation of the principle of polarity, but it does serve as a general model that has practical applications.

There are parts of the *ƛuukʷaana* that are not frightening. For example, there is the part where the speaker for the *Ḥawʼił*, or chief, may display and fill the big house with grand oratory and sacred chanting, which is immedi-ately echoed by a female elder in the audience. The female's response takes the form of spiritual chanting that signifies approval; it is similar to when, in a temporal setting, an English audience may approvingly shout "hear! hear!" to indicate synchrony and agreement with a given speech. Each is a form of witness, a form of agreement and affirmation. The response from the audience functions to say: "I hear you and understand what you say, and it is true!"

After much time spent in serious ceremony, there comes a time of delib-erate fun. The late Tseshaht artist, actor, and author George Clutesi called this time *wikiičił,* a time when all roles and responsibilities are forgotten, meaning that everyone is at the same level and that no one is superior to or more important than anyone else. *Wikiičił* is a time of laughter and merriment, without reason or cause – except that the spontaneous merri-ment is a purposeful ceremony intended to foster the health and well-being of community. There is irony here because, although *wikiičił* is a time of *merriment without a cause,* it is also a time that is intentionally set aside; therefore, it represents a form of managing the collective human psyche. It is a part of a governance system that, constitutionally, sets aside a time for laughter. Ultimately, after a period of days or even longer, the lost children are rescued by warriors and reunited with family, community wholeness is restored, and the daily struggle is resumed.

The foregoing describes some of my early experiences, which owe nothing to European or missionary influences and owe everything to the legacy of my ancestors. I experienced other teachings and practices that had to do with family, but these have now declined to the point of extinction. One such practice was called *ʕinaak* (a nurturing, cooing sound), and it was universally observed by older women, mothers, and grandmothers. Its full expression cannot be captured in words as it involved voicing a sound that expressed pure feeling and emotion. This sound was extended, melodic, and unbroken, and it was meant to express endearment, often to a baby but sometimes to a toddler and even to a young person. It was a warm sound of relational acceptance. Even now, to think, reflect, and meditate upon this experience brings a deep longing for a lost place of refuge, for the lost medicine to soothe troubled souls.

A social practice related to *ʕinaak* was the common greeting. Every time relatives came to our house, their visit was treated like a very special meeting with special people. As a little child, I would simply stare at these people (whom I had never seen before since they lived in another community up the coast from us), and Nan would say to me in our language: "They are your relatives! Always greet your relatives enthusiastically." This teaching involved a self-conscious application of the meaning and purpose of *ʕinaak,* an affirmation that the quality of human relationships is of primary importance.

In fact, when the stories about Wolf, Deer, and Bear were told together with the stories about how to take down a tree to make a canoe, it was quite evident that the quality of relationships between life forms was of universal importance. In the Nuu-chah-nulth language, when we speak of the importance of relationships, we also speak of relationships with Wolf, Bear, Deer, and, indeed, with all other life forms for they are all *quuʔas,* just like us. Roy Haiyupis, an uncle of mine who served as an elder to the Scientific

Panel for Sustainable Forest Practices in Clayoquot Sound, told the story of when, as a boy, his grandfather showed him how to take down a tree. Roy described how his grandfather revered even the bushes and foliage around the tree and treated each with respect. When this paying of respect to the bushes and foliage was completed, his grandfather lifted up his voice to communicate with the great tree. The African shaman Mutwa affirms that such communication was also common to his people when he says: "But we humans can communicate with all the animals, and with plants also."[8]

My Aunt Trudy's husband tells another story that illustrates a common teaching concerning resources. When he was about thirteen years old, he had gone to Tofino inlet by himself to hunt ducks. He encountered so many ducks that, when they rose up to fly, they darkened the sky above him. He took no aim but simply pointed his gun into the sky and pulled the trigger. When the sky cleared, there were sixty-five ducks floating on the water. When he arrived at the nearest village, he was met by an elder who was a relative and was instantly chastised: "Never take more than you need!" he was told. "Now hurry and find some relatives and give them some of your ducks." There are many teachings like this, but, like any laws in any country, they are not always followed. My aunt's husband was a young boy who also lived in changing times, so perhaps he cannot be faulted for his breach of an ancient teaching.

In time, I was sent to a residential boarding school, where I spent twelve years. Then I went to university, got married, and had two boys, who, in turn, got married and now have six children between them. Today, my oldest grandchild has begun his third decade of life, while the youngest grandchild has just begun his first. The inexorable and fast-paced march of technological civilization, with its cell phones, televisions, radios, computers, music videos, endless CDs, and single-family housing covered in plastic on the outside and wall-to-wall carpeting on the inside, has come to my little village. More than 50 percent of the community's members live in urban areas like Port Alberni, Nanaimo, Victoria, Vancouver, and even farther away, whether for reasons of employment, education, or whatever. After several visits to my community over the years, it strikes me now that the sound of ʔinaak has been overshadowed by the sound of technology. I do not hear ʔinaak anymore.

After my early traditional training within my culture, the residential school system immediately began to undo my prior education. On the first day, as I walked out of the red-brick building, the first person whom I spotted on the playground was my cousin. I was so happy to see someone familiar that I shouted to him in our language, "Friend, where are you going?" Immediately, a first-floor window flew open and a middle-aged woman glared down at me and warned: "Don't you speak your language ever again – or else!" It seemed that, up until then, everything I had learned and experienced was good and nurturing, but suddenly everything I had learned had become evil and unacceptable.

Since my father died, circumstances dictated that I had to attend residential school. Slowly, I learned the new ways. Eventually, my long journey into a world different from the one in which I was born culminated in a doctoral degree. Up to the time of this achievement, the colonial myths propagated about Aboriginals were still commonly held. In 1969, when I began to teach elementary school, I was married with one son and another on the way. During this period, if I wanted to stay in a motel or a hotel it was routine for desk clerks to say that they had no rooms available. My brother Luke told me that he had applied for a job advertised in a local sawmill. After he was rejected, with the assurance that the job was already filled, a white friend who was with him suggested that they wait outside the hiring office. Sure enough, a young white man entered the office soon afterward and, when he came out and was asked if he had been hired for the job, he said, "Yes!" For the most part, the multitude of negative stereotypes about Aboriginals created during the colonial era characterized Canada's relations with its indigenous populations for the entire twentieth century.

Since my own early experiences contradicted these stereotypes, I began to ask myself how it was that my ancestors came up with, or created, their way of life. Where did they get their ideas and information, and how did they arrive at their system of teachings, their practices? This line of inquiry leads to what I call *cultural assumptions*. No one wakes up in the morning and says: "According to my ontological belief system, according to the worldview developed by my ancestors, I will now give thanks to K̓ʷaaʔuuc for another day and also ask for strength and guidance." Rather, one simply takes a moment to lift up the heart in thanksgiving and petition, and then the day begins.

One place of entry into the knowledge foundations of the Nuu-chah-nulth is Carobeth Laird's book *Mirror and Pattern: George Laird's World of Chemehuevi Mythology.* What she writes of the Chemehuevi is also true of the Nuu-chah-nulth: *The myths, therefore, were the pattern and are for us* [Western observers] *the mirror of a culture.*[9] Myths, when interpreted from a contemporary standpoint, are stories without any basis in scientific truth. Yet, to see that myths can be interpreted as a "pattern" of culture is to reach a conclusion similar to Huntington's: "World views and causal theories are guides to understanding international politics."[10] Jean Shinoda Bolen is more specific when, in reflecting upon Western civilization, she writes: "The mythology of a culture ... instructs us about the values, patterns, and assumptions on which ... culture is based."[11] The bases of a culture are the direct result of the way reality is viewed, of one's worldview. Hence they are guides to our understanding about the way people live.

Although myths appear to focus upon the non-physical or spiritual, they comprise stories that are meant to be practical. They are meant as guides to understanding the nature of reality. In this sense, myths are not necessarily in opposition to the intent of scientific inquiry; rather, they simply take a different route and employ different terminology. Whereas scientific inquiry depends on theory, indigenous knowledge systems depend on myths. Both theory and myths can be tested. The Nuu-chah-nulth test is called the *ʔuusumč,* the vision quest. It can truly be said that, for the Nuu-chah-nulth, as a research process the *ʔuusumč* is rooted in, and validated by, lived-experience. Consequently, when *ʔuusumč* was practised universally, as it was in precontact Nuu-chah-nulth society, it could unveil truths about the nature of reality that were so reliable that they sustained a way of life over millennia. If the foundation for a perspective on reality is untrue, then there is no possibility that it can have a practical application. What people believe and practice must have enough practical substance that it enables them to survive, to live adequately, and even to thrive.

Just as scientific "truths" about the workings of nature are based on systematic theory and research, so, too, indigenous "truths" are based on stories and on the "research" that is conducted during the *ʔuusumč.* Carobeth

Laird's insight that myths form a pattern that mirrors a culture is reflected in the table below. A parallel statement can be made about Carobeth Laird's Western culture: "The scientific theories,[12] therefore, were the pattern and are for us [indigenous peoples, in this case] the mirror of a culture."

COMPARISON OF SCIENTIFIC AND INDIGENOUS KNOWLEDGE SYSTEMS

		Test of theory	Results reported	Applications
Scientific way	Theory	Scientific method	Papers, conferences, books	A world of high technology
Indigenous way	Origin stories	Vision quest method	To each family at feasts, potlatches	A world that strives to balance and harmonize

In *Tsawalk* I suggest that myths – or origin stories, as I prefer to call them – serve the same purpose as does scientific theory since they provide insight into the nature of existence. The test of theory for an origin story is, again, the *ʔuusumč*. The experiences of the *ʔuusumč* were individual and personal, but the practice was universal. Every household necessarily had practitioners of the *ʔuusumč* because it was firmly believed that it was not possible to succeed in life without it. Not only did *ʔuusumč* reveal spiritual powers, spiritual gifts, songs, directions, and information, but its application also culminated in successful hunts for food, which, today, would be classified as successful research outcomes.

The many types of traditional feasts were the venues for public reporting about the findings of the *ʔuusumč*. The reporting was conducted not with papers or PowerPoint slides but, rather, with rattles, chants, songs, dances, regalia, and demonstrations of spiritual powers and medicinal gifts. And the reporting was often conducted by well-trained orators. Unlike the process of contemporary science, which is exclusive to a relative few in each discipline (e.g., the few who study and practise physics or the various branches of chemistry, biology, etc.), the ancient Nuu-chah-nulth process of knowledge and power production was practised by members of every household. As a consequence, at any given feast, where people made various kinds of power or knowledge claims, those claims could be judged by an experienced and knowledgeable audience.

From a Nuu-chah-nulth perspective, myths are a reflection of, and a source of wisdom about, the nature of reality. They are also of mysterious origin. Yet, it can be deduced from the meaning of the name of Ḵʷaaʔuuc, Owner of All, that they may well be the ultimate source of Nuu-chah-nulth knowledge. Myths originate with Ḵʷaaʔuuc. This is something believed but it is also something that can be neither proved nor disproved with existing research tools. This belief is no less sound than are current scientific "beliefs" about string theory, parallel universes, and other theories about the nature of reality that cannot be proved or disproved with existing research tools.

To conclude this section, whatever the source of myths, or origin stories, they have been most reliable in providing guidance to the ancient Nuu-chah-nulth regarding how reality works and how people might successfully manage within this reality. Although their guiding worldview did not always enable the Nuu-chah-nulth to achieve an ideal balance and harmony among and between all life forms, it did enable them to enjoy enough success to convince them that their stories were reliable and valid. For the most part, their belief system and their cultural practices provided the Nuu-chah-nulth with the basic necessities of life. More important, a constellation of teachings was developed to maintain and to enhance life's major purpose, namely, the development of harmonious relationships between and among all life forms. These teachings applied to every category and dimension of life, whether within an extended family, between nations, or between all other life forms, including those forms (such as plants and animals) that are now considered to be simply resources. These teachings also applied to the relationship between the physical and the spiritual realms.

In my view, the ancient Nuu-chah-nulth were able to exist because they had stories that taught them not only about the world in which they lived but also about themselves. Over time, they became more successful at exploring this knowledge and translating it into practice. This success, qualified by the limitations of human nature, meant that they were able to increase the quality of their lives. It is unlikely that the source of this knowledge and practice came exclusively from the human psyche and intellect since these are potentially interconnected with the life of the spiritual domain. If Ǩʷaaʔuuc is the Owner of All, then Ǩʷaaʔuuc must also be the source of life and, according to Nuu-chah-nulth belief, must play a significant role in the development of human lifeways. This perspective provides a sound explanation of why, although in the twenty-first century the Nuu-chah-nulth way of life has emerged as a broken cultural expression, the Nuu-chah-nulth have not been completely destroyed – any more than, in an earlier age, the ancient community of Ahous was completely destroyed by *Aɬmaquuʔas* (Pitch Woman).

TRANSFORMATION
CLEESTERK '11

RAVEN BECOMING
INSIGNIFICANT LEAF

3

Genesis of Global Crisis
and a Theory of Tsawalk

Samuel Huntington has observed: "The great political ideologies of the twentieth century include liberalism, socialism, anarchism, corporatism, Marxism, communism, social democracy, conservatism, nationalism, fascism, and Christian democracy. They all share one thing in common: they are products of Western civilization."[1] These ideological products of Western civilization also have in common a natural polarity, evident in the opposition, for example, between liberalism and fascism and between Christian democracy and communism. Another feature of these Western products, which is held in common by societies the world over, and to which indigenous peoples of the Americas can testify, is the use of warfare in an attempt to eliminate opposition, which has been true not only in human conflict but also in the way nature has been devastated by human activity. If Huntington is correct, then the implication is that the current stage of global crisis must be rooted in these "great political ideologies of the twentieth century,"[2] which are reflected in the current world order.

However, according to the theory of *tsawalk* (one), any planetary stage of crisis must, by definition, be a shared responsibility, a shared experience. The Nuu-chah-nulth notion that reality is fundamentally an interconnected and interrelated whole regardless of its seeming polarity, its seeming contradictions, has been interpreted historically in one of two ways. The first is indicated by what Huntington defines as the progress of civilization, which, it is argued, sets the current stage of crisis. This crisis is defined as one of interrelationships between humans and between humans and other life forms. In this interpretation of reality, the practice has been to eliminate opposition to, for example, the imagined ideal of progress. Until recently, this has meant the extirpation of non-European beliefs and lifeways through legislation and policy. This has involved a process of resource extraction that has proven degrading to nature. The current global environmental movement is one response to the latter process, and it has had some measure of success (e.g., in the creation of both land and marine preserves that prohibit any kind of exploitation of natural habitat).

The second interpretation of the Nuu-chah-nulth notion of a unified but polarized reality remains largely peripheral to centres of power and influence. In this interpretation the focus is on the development of sustainable relationships between life forms. In other words, it is on managing polarity by working to transform the inherent contradictions of reality into a sustainable balance and harmony so that all life forms can continue to live. This interpretation of the interrelated characteristic of reality seeks not to eliminate opposition but, rather, to employ the natural oppositions and apparent contradictions of reality to realize wholeness.

If the theory of *tsawalk* has some credence, as I believe it does, then the responsibility for the current planetary crisis must be shared by all: both those who colonized and those who have been colonized. The idea of shared responsibility must be understood within the context of an inherent polarity that includes its opposite – the refusal to share responsibility. To this day, responsibility for the planet has not been shared. To the contrary, for the past five hundred years the colonizers have put all responsibility on the colonized.

In support of my proposition that life on earth is a shared responsibility, I turn now to the life of that great Mongol leader, Genghis Khan, whose conquests had a purpose that touches on this notion of shared responsibility. For this section I build upon the ideas of Jack Weatherford, a renowned anthropologist who wrote about the historical significance of the conquests of Genghis Khan.[3] It seems that Roger Bacon, who lived in the thirteenth century, may have been the first to acknowledge the contributions the Khan's conquests made to the development of the modern age. Bacon recognized not only that the Mongols enjoyed *martial superiority* over the rest of the world but also that they succeeded *by means of science.* Four centuries later, in 1620, as Weatherford records, Sir Francis Bacon affirmed these observations about this great Mongol Khan:

> He designated printing, gunpowder, and the compass as three technological innovations on which the modern world was built. Although they were "unknown to the ancients … these three have changed the appearance and state of the whole world; first in literature, then in warfare, and lastly in navigation." More important than the innovations themselves, from them "innumerable changes have been thence derived." In a clear recognition of their importance he wrote "that no empire, sect, or star, appears to have exercised a greater power and influence on human affairs than these mechanical discoveries." *All of them had been spread to the West during the era of the Mongol empire.*[4]

It will be noted that my emphasis in the above quote is not on the three technological innovations of printing, gunpowder, and the compass but, rather, on the equally important process of knowledge-transfer and knowledge-sharing. This process is central to the principals of *tsawalk* because it is central to creating balance and harmony in a conflicted environment. However, in Genghis Khan's era, it seems that, in order to clear the way for these contributions to a new world order, it was first necessary to destroy the old world. Indeed, the Great Khan implied as much when he said to the vanquished elite of Bukhara: "If you had not committed great sins, God would not have sent a punishment like me upon you."[5] This notion of warfare as punishment was not unique to Genghis Khan; rather, during his time, it was a common belief across many cultures and societies. One example of this practice may be found in my own ancestral community of Ahous. As late as two hundred years ago the ancient Nuu-chah-nulth engaged in warfare with their neighbour Otsoo because of a perceived violation of beliefs considered essential to life. Today, in many societies, warfare may now be understood to represent an older order of creation – one that, hopefully, is giving way to another order, an order in which polarity is considered a benefit rather than a threat, a challenge to grow rather than to destroy.

It is one thing for Weatherford to claim that Genghis Khan contributed to the creation of the modern world and quite another to acknowledge its truth. On this point he writes: "With the passage of centuries, scholars weighed the atrocities and aggression committed by men such as Alexander, Caesar, Charlemagne, or Napoleon against their accomplishments ... For Genghis Khan and the Mongols, however, their achievements lay forgotten, while their alleged crimes and brutality became magnified."[6] It is these Mongol achievements, he argues, that allowed the stage to be set for the Enlightenment, which, in turn, helped to create the modern world. The fact that these Mongol achievements were lost to history is no doubt why later historians make no note of them. I. Bernard Cohen, who was, until he died in 2003, a professor of the history of science at Harvard University, believed that "Enlightenment philosophers derived their ideas of freedom from Newtonian physics"; however, Charels C. Mann notes that "a plain reading of their texts shows that Locke, Hume, Rousseau, and Thomas Paine took many of their illustrations of liberty from native examples. So did the Boston colonists who held their anti-British Tea Party dressed as 'Mohawks.'"[7]

Before the histories of the Americas influenced the scholars of Europe, the latter were influenced by the conquests of Genghis Khan. From the latter historical perspective: "In the end, Europe suffered the least yet acquired all the advantages of contact ... The new technology, knowledge, and commercial wealth created the Renaissance in which Europe rediscovered some of its prior culture, but more importantly, absorbed the technology for printing, firearms, the compass, and the abacus from the East."[8] Europe's acquisition of "all the advantages of contact" through the transfer of knowledge that resulted from the conquests of Genghis Khan eventually led to the development of the "great political ideologies" of Western civilization outlined by Huntington. While these claims of Mongol influence on the West may be contested, what may not be contested is the fact of contact (which, by definition, constitutes an influence). I do not claim either that the Mongols were superior human beings or that Western accomplishments were derived solely from Mongol influence; rather, I claim that historical events are interrelated and interconnected.

In what specific ways, then, did Genghis Khan contribute to the development of the current world order? Weatherford is emphatic that the Mongols can claim no inventions of their own:

> The Mongols made no technological breakthroughs, founded no new religions, wrote few books or dramas, and gave the world no new crops or methods of agriculture. Their own craftsmen could not weave cloth, cast metal, make pottery, or even bake bread. They manufactured neither porcelain nor pottery, painted no pictures, and built no buildings. Yet, as their army conquered culture after culture, they collected and passed all of these skills from one civilization to the next. The only permanent structures Genghis Khan erected were bridges.[9]

Bridge building is an apt metaphor to describe Mongol contributions. It is a metaphor that supports an indigenous worldview characterized by interrelationships and interconnections, which is the theme of my work. Genghis Khan built bridges between peoples who were divided by exclusionary beliefs, ideologies, and customs. He and his successors did this, Weatherford explains,[10] through the introduction of paper currency intended to be used everywhere. The Khan created roads and bridges to facilitate international trade and attempted to unite the world of that time with a single international law and a universal alphabet. He also strove to create universal basic education for all children so that everyone might become literate. Calendars were refined and combined to create a ten-thousand-year version that was more accurate than any previous calendar, and numerous maps were made available so that the world could be more easily navigated. He created an international mail system. He brought together representatives of three major religions – Christians, Muslims, and Buddhists – and attempted to reconcile their differences. This last effort failed, as such efforts continue to fail today. However, what was not a failure was the outcome of the conquest itself. Once the conquest was complete there then began an unprecedented rise in free trade and cultural enrichment. Unfortunately, this state of affairs did not outlive the Khan, whose empire began to crumble soon after his death. However, his vision of an interrelated global community remains.

Interestingly, the Mongols were similar to Nuu-chah-nulth and other Aboriginal peoples in that their method of knowledge acquisition, the view of reality that guided their social, economic, and political life, fully integrated the practical demands of living with shamanic insight.[11] Genghis Khan employed a ʔuusumč method that successfully integrated the non-physical with the physical, which, in this case, meant doing away with an apparently corrupt world order so that a foundation might be laid for a better one. The chroniclers of Armenia, including those in the trading cities of ancient Russia, recorded their understanding of the appearance of Genghis Khan and his armies as a "punishment from God."[12] While warfare is no longer considered an acceptable solution to problems, a significant result of Genghis Khan's conquest was that "Europe suffered the least yet acquired all the advantages of contact."[13]

This story about Genghis Khan provides one example of how to integrate different dimensions of reality. For Genghis Khan, myths, or stories about creation, provided an orientation towards reality that provided a worldview that guided him in all things. These myths are the same kind that, for millennia, guided Nuu-chah-nulth peoples.

At the beginning of the colonial era, the "Old World" was energized by stories of untold wealth in the Americas. Then, as now, the primary motivation for conquest was material possessions. Genghis Khan sought material possessions, but only through the indigenous method, which is guided by shamanic powers. In the words of the great religions of the world, he was guided by the hand of God, a point debatable within Western scholarship but not within indigenous knowledge systems. Therefore, although some may dispute the claim that one can be guided by the hand of God, there is a significant group of indigenous peoples, around 200 million of them, whose traditional way of life has adhered to the integrative methodology employed by Genghis Khan.

Is it possible to reconsider the utility of the integrative methodology that drove the conquests of Genghis Khan and that allowed each Nuu-chah-nulth household to develop a sustainable way of life? Is it possible to consider a way to negotiate our reality (Western and Other), with all of its apparent contradictions, and envision an alternative to conflict and exclusion, finally achieving a vision of harmony and inclusion, a vision of continuity for all of life on earth, on Haw'iɫume? From a Nuu-chah-nulth perspective and lived-experience, there is reason to say yes.

The experience of my great-grandfather Keesta during the *ʔuusumč* he undertook to ensure a successful whaling expedition is one example of a Nuu-chah-nulth approach to the possibility of balance and harmony. During an extended and challenging spiritual quest up in a local mountain, Keesta said to himself that he would know that he had a positive answer from the spiritual realm when the curly tail feather of a mallard duck, which he took with him, straightened out of its own accord.[14] In terms of empirical verification, then, for Keesta, the straightened feather would be one example. A successful whaling expedition would be a second. By themselves, these two examples of empirical verification are insufficient to validate a one-time event, but they are more than sufficient when affirmed by similar experiences shared by members of every Nuu-chah-nulth household. As may be noted countless times by other practitioners, Keesta's type of experience during a *ʔuusumč* was not unusual. It may be said, then, that Keesta had at least three ways to empirically verify the authenticity of his spiritual quest: the feather that straightened out of its own accord, a successful whaling venture, and the knowledge of similar kinds of spiritual quests performed by members of every Nuu-chah-nulth household.

There seems a genuine, if incipient, interest in the pragmatics of an integrative methodology of the non-physical and physical domains from some contemporary non-indigenous scholars. For example, Graham Saayman, a trained psychologist from the University of Natal in South Africa and the universities of McMaster and London in Canada, has recently published a book with the telling title *Hunting with the Heart: A Vision Quest to Spiritual Emergence*. During his research on whales and dolphins he describes some of his unusual experiences with animals while sitting on a bluff in the hot African sun waiting for the cetaceans to appear. He explains how these unusual experiences came about: "the less I moved the smaller and less exposed I would become and the less I would suffer from sunburn."[15] During this intentional and personal condition of *inner smallness* – otherwise known in the *ʔuusumč* tradition as the necessary humble condition required for a spiritual process – a snake fell asleep at his feet and a rarely seen, tiny cliff-dwelling gazelle, known in South Africa as a Klipspringer, appeared to him. Graham writes: "This delightful little doe studied you silently as part of the panorama for some long minutes and then it would be gone again, as if it had vaporized into the ether."[16] These sorts of descriptions are usually dismissed by scientists, but they are exciting from an indigenous perspective. We know that one necessary condition for a successful ʔuusumč is to maintain a deliberate state of inner smallness. Had Saayman actually been on a ʔuusumč, the stage might have been set for a possible communication with the spiritual realm, which may have been effected through the Klipspringer.

The foregoing examples of spiritually oriented experiences may have significant parallels with scientific experimentation in the invisible world of protons, neutrons, and electrons. In both the methodology of the ʔuusumč and the methodology of physics, unseen forces can be detected indirectly. The Big Bang theory of the origin of the universe is based on a method of indirect detection. The current rate of cosmic expansion indicates an original explosion for which there is no direct evidence.

The method whereby Genghis Khan integrated metaphysical powers into practical human affairs, which resulted in his conquests, provided a model that may have eventually paved the way to globalization. If it is true, as some Western scholars would have it, that Genghis Khan conquered for the sole purpose of destroying peoples and civilizations, then it is passing strange that Marco Polo should risk his life to go on a pilgrimage to seek the court of the Khan's grandson, Kublai Khan. Would Marco Polo, or any other explorer, travel halfway around the world to visit a destroyer of peoples and civilizations?

His friends nicknamed Marco Polo *Il Millione,* in a reference to his tall tales,[17] but this nickname did not prevent his entry into the historical record, which is fitting, as history consists of stories made legitimate. In the second half of the thirteenth century, Marco Polo reached the court of Kublai Khan in China, and it was Marco Polo who brought back to Europe the stories of fabulous riches and wealth to be found there. Although we who live in the Western world have all heard about Marco Polo, we have not learned about other pilgrims who travelled from the East. At about the same time as Marco Polo's adventures, two monks from China, Bar Sauma and Mar Markos, went on a pilgrimage to Jerusalem.[18] Western missionaries had entered China several centuries ahead of Marco Polo, and it was Mar Markos, a Christian Chinese monk, who convinced his friend Bar Sauma to travel with him on this pilgrimage. Bar Sauma and Mar Markos were received by the kings of France and Byzantium, and, in addition, they petitioned the king of England to form an alliance with Kublai Khan in order to capture Jerusalem, which was then under the rule of the Muslim Mamluks. In the end, what with all of this diplomatic activity, they did not make it to Jerusalem. But what this account suggests is that the natural process of globalization began centuries before the time of Genghis Khan, back in the time of the arrival in China of the first missionaries.

The pilgrims from China seemed to have had a very different agenda from that of Marco Polo. Both pilgrims sought wealth, one material and the other non-material. However, during the past five hundred years, it has been those seeking material wealth who have prevailed. It was a world order based on acquiring material wealth that turned the Nuu-chah-nulth world order upside down and led to the world we have today. Although the exploits of Genghis Khan may have been forgotten, it was news of his grand courts in China that motivated Christopher Columbus to sail west in 1492.

In an attempt to see more clearly the reasons for the explorations to the Far East set off by the stories told by Marco Polo, it might be useful to consider the general state of the peoples of Europe during this period. Ronald Wright, a novelist, historian, and essayist, who has been published in sixteen languages, provides a glimpse into the general condition of Europe when Columbus set sail on his quest to find a trade route to the Far East:

Europe in 1492 was a small affair. The British Isles had only 5 million people, Spain about eight. Political boundaries were essentially those which had resulted from barbarian migrations after the fall of Rome. The Franks had settled in France, the Germani in Germany, the Angles and Saxons in England, the Vandals and Visigoths in Spain. These patterns have altered remarkably little from the seventh century to the twentieth, though frequent attempts to alter them have made the soil of Europe among the most bloodstained on earth.

European secular government was a tangle of decayed feudal loyalties and personal ambition. The last proper roads had been built by the Romans more than a thousand years before. The rapidly growing cities were unplanned, ramshackle, without sanitation, seething with poverty and disease. If famine struck a region, the state was quite unable to provide relief. Life expectancy oscillated between the high teens and low thirties, lower than in the most deprived nations of today. The achievements of Europe were technological, not social. It had the best ships, the best steel, the best guns; it also had conditions desperate enough to make its people want to leave and use these things to plunder others.[19]

Some 260 years after Columbus sailed to the Americas, Jean-Jacques Rousseau observed: "Man was born free, and he is everywhere in chains."[20] It seems that Rousseau's Europe had not improved from the time of Columbus, remaining "a tangle of decayed feudal loyalties ... rapidly growing cities ... without sanitation, seething with poverty and disease." This is quite a different picture of Europe from that which I received in elementary school, where I was imbued with the grand impressions of a refined civilization. Much later I also learned that, early in the seventeenth century, some Huron and Montagnais (also called Innu) returned from France to report to their chiefs on what, to them, were the astonishing general conditions typified by the "needy and beggars." In the words of the Jesuit missionary Sagard, the Huron and Montagnais chiefs chastised the French missionaries and told them that "if [they] had some intelligence [they] would set some order in the matter, the remedies being simple."[21]

Not only was the Europe of that day a small affair and its people "everywhere in chains," but the majority were illiterate. Consequently, it might be imagined that such unfavourable living conditions provided the necessary motivation for people like Columbus to set sail into uncharted waters and territories. During the settlement era, many people fled to the New World because, in addition to Rousseau's chains, they suffered from religious persecution and severe inequalities. Under these kinds of political and social circumstances it may be reasonable to ask if the lived-experiences of oppression could lead naturally to the Enlightenment. The answer must be yes. And the follow-up question must be: "How?" Could it be that the discovery of the New World meant not only the contested acquisition of natural resources but also the acquisition of new and exciting ideas about an egalitarian lifestyle – a lifestyle hitherto unknown to the Old World?

Amerigo Vespucci

Amerigo Vespucci was a contemporary and an acquaintance of Christopher Columbus. There is a dispute about how America came to be so named, but what is not in dispute is that it was named after a European. Although history teaches that America is named after Vespucci, its discovery is credited to Columbus. In elementary school we learned that Columbus bumped into America while he was looking for a trade route to the Far East. Whereas Columbus thought he had landed in the Far East, Vespucci disagreed because he had also sailed across the same ocean in search of the same trade route. Vespucci said to Columbus:

> With all the deference that is due to your Excellency's superior wisdom and experience, I would state, that therein lies the very point of our difference. I deem it by no means certain that your ships have touched the territories of the Grand Khan at all, but rather a land which has hitherto been alike unknown to him and to us. Thousands of leagues may yet intervene between that land and his dominions, whether of sea or earth remains to be discovered; and I judge in this wise, as well from the accounts of cosmographers, who have written upon the subject, as from the description of the barbarous natives, which you yourself have fallen in with, in your recent discoveries. The accounts of those who have penetrated to the distant regions of the East, lead us to understand that the subjects of the Grand Khan live in the midst of the most profuse wealth and luxury, and bedeck themselves with superfine garments, and much gold and jewelry. These people, however, are naked and wild, and may

be looked upon as little superior to the beasts, and I think cannot be in any way connected with a monarch of such magnificence.[22]

Columbus replied, "Nay, nay, good Signor Vespucci, I have the confidence in my heart that, you are mistaken."[23]

In hindsight, we know that Vespucci was right and Columbus wrong. Just as Vespucci is comparatively unknown as an explorer, so also is his contribution to the propagation of new ideas into Europe. The reason for this is that his letters to Lorenzo di Pier Francisco Medici (describing his voyages to the New World) were printed after the latter's death and were apparently altered (with the introduction of some exaggerations) in order to make them more interesting and saleable. What is not contested, however, are his descriptions of the New World and its inhabitants, and that is because these observations were endlessly confirmed by other navigators and observers. "They are not accustomed," he said of the New World Natives, "to have any Captain, nor do they go in any ordered array, for everyone is lord of himself ... these people have neither King, nor Lord, nor do they yield obedience to any one, for they live in their own liberty."[24]

It might be said that to be "lord" of oneself and to live in one's "own liberty" are descriptions of the highest form of being, the highest form of enlightenment. Vespucci and others observed human societies in which people were neither obsequious nor afraid to move about freely. This social condition was seen to be remarkable. Unlike European societies of that day, where people were "everywhere in chains," subjected to serfdom, subjected to various lords, it appeared to Vespucci that, in indigenous American societies, people came and went without restraint. This was a predominant characteristic of New World societies, even though many of them, such as the Nuu-chah-nulth, had slaves in addition to their otherwise egalitarian practices (slavery, of course, was not uncommon to European democracies).

What enabled Native peoples to appear to Vespucci as being lords of themselves was not a matter of instinct and primeval urges, as many have imagined, but, rather, of continuous disciplined, thoughtful development that, in the prime of life, often involved the human spirit entering a "trial by fire" during a ʔuusumč. Vespucci observed the behavioural outcome of a human development process that – unlike the contemporary human development process, which has become institutionalized in the form of a public education system – was often conducted in secret by each family. In reference to the letters of Vespucci, Jack Weatherford remarks in his book entitled *Indian Givers* that, as "the first reports of this new place filtered into Europe, they provoked much philosophical and political writing."[25] It is these writings that emphasized the "liberty of the Indians and the virtual enslavement of the Europeans." This "became a lifelong concern for Rousseau and eventually led to the publication of his best-known work, *Discourse on the Origins of Inequality,* in 1754."[26]

The great irony is that the liberty that the American Aboriginal enjoyed would soon be subsumed by a colonizing project that meant a loss not only of freedom of movement but also of thought, to speak one's own language, and sometimes to live. Vespucci's interpretation of Aboriginality, later greatly elaborated upon by the thinkers of the Enlightenment, created a breach between the peoples of the Americas and the peoples of Europe that would not be mended. It is this breach that made it so easy to manufacture a convenient "other" to demonize, exclude, and extirpate. It is the same breach that was used to justify the scientific belief of the day, which held that this other was fated to succumb to the genocidal process of natural selection. Until the early twentieth century, indigenous peoples were thought of as a vanishing race.

It must give one pause to consider the credulity of European beliefs about Aboriginal peoples. On the one hand, there is the belief that the indigenous peoples of an entire continent were constantly at war with one another (as Vespucci, Daniel Defoe, and others indicated), exercised the utmost cruelty, ate one another, and generally lived without laws and morals (all the while somehow maintaining their population bases). On the other hand there is the belief that indigenous peoples lived some kind of utopian existence. Still, Vespucci's observations of the indigenous peoples that he encountered have considerable value to me because they provide an opportunity to present an alternative interpretation.

"Warfare," Vespucci wrote, "is used amongst them, against people not of their own language, very cruelly, without granting life to any one, except [to reserve one] for greater suffering. When they go to war, they take their women with them not that these may fight, but because they carry behind them their worldly goods: for a woman carries on her back for thirty or forty leagues a load which no man could bear: as we have many times seen them do."[27] Since Vespucci's letters were "much-discussed" throughout Europe,[28] there seems little doubt that it was from them that the European public first learned about the New World and its peoples. The information was new and generally accepted without question. As a result, the notion that indigenous peoples were warlike became a belief of long standing.

Much more damaging to indigenous peoples everywhere were some of Vespucci's gross exaggerations, as in the following:

> They are so inhuman that they outdo every custom [even] of beasts: for they eat all their enemies whom they kill or capture, as well females as males, with so much savagery, that [merely] to relate it appears a horrible thing: how much more so to see it, as, infinite times and in many places, it was my hap to see it: and they wondered to hear us say that we did not eat our enemies: and this your Magnificence may take for certain, that their other barbarous customs are such that expression is too weak for the reality.[29]

It is not physically possible for Vespucci to have witnessed an "infinite" number of instances of cannibalism, nor could he have witnessed these ubiquitously. Perhaps this is one of the ways in which his letters were altered and exaggerated in order to make them more saleable. Cannibalism is a fact of ancient history, although its degree of practice is in dispute. Today, all peoples have presumably given up cannibalism, yet it must be realized that, even in ancient times, this practice was not conducted instinctively or without a rationale. And it was not common practice to eat just any human for it was believed that what was eaten directly affected the whole of one's life. My grandfather once told me that eating wild ducks would make me weak because – he did not have to explain – ducks are weak beings. The goal of a warrior and hunter was to be strong so that he could provide for family and community. So, unlike eating ducks, eating the heart of a lion would involve partaking of its strength and perceived valiant spirit.

From a Nuu-chah-nulth perspective, the entire universe is replete with powers, both creative and destructive, and these powers can be harnessed. When is the best time to go on a vision quest, a *ʔuusumč?* When the moon is waxing and getting bigger and stronger, not when it is waning and getting smaller and weaker. The best time to be in the ocean for ritual purposes is when the tide is rising and becoming full, not when it is falling. By observing these protocols, each vision seeker can strengthen him- or herself. In many different ways, indigenous peoples of old sought to maximize their strength. A pregnant Nuu-chah-nulth woman was taught to stay away from angry people and from funerals so that the foetus would not be negatively affected and weakened. Babies would begin their training while in the womb, when the mother or grandmother sang to them. They would begin life to the sound of lullabies that oriented them to strengths of Nuu-chah-nulth lifeways. If Vespucci had been a more insightful observer, he might have learned that, at any given time and in any given era, some indigenous peoples regularly engaged in limited warfare (or, more appropriately, periodic skirmishes), while others avoided warfare altogether.

As Vespucci travelled along the coast of this new land, he wrote: "We departed thence, and made our way to the inner part of the bay: where we found such a multitude of people, that it was marvellous: with whom on landing we made a friendship: and many of us went with them to their villages, very safely, and well-received."[30] Of the land of these same people, Vespucci notes: "The soil abound greatly with everything they need for subsistence, and the people [were] of polite behaviour and the most pacific we had as yet met with."[31]

From a European perspective, "polite behaviour" may mean a number of things, depending on one's orientation. Among equals, polite behaviour is a sign that one is well brought up, educated, civilized, and usually a gentleman or gentlewoman. Between unequals, polite behaviour on the part of the inferior person is considered obsequious and appropriate. From a Nuu-chah-nulth perspective (as one example of an indigenous perspective), polite behaviour is the social consequence of being taught to be

ʔaapḥii (friendly). ʔaapḥii is closely related to iisʔaḵ, which refers to a respect that "necessitates a consciousness that creation has a common origin."[32] Consequently, friendliness was not just a kind of behaviour that someone decided was appropriate for humans; rather, it was a way of acknowledging Ḵʷaaʔuuc (Owner of All – Creator). For the Nuu-chah-nulth people, polite behaviour was about more than being well brought up, well educated, or well situated in terms of social status; it was also about a way of being human. Physical growth and development was integrated with spiritual growth and development.

At still another peaceful village, Vespucci reported that those of his party were neither attacked nor ill-treated but, rather,

> were received with such and so many barbarous ceremonies that the pen suffices not to write them down: for there were dances, and songs, and lamentations mingled with rejoicing, and great quantities of food: and there we remained the night: where they offered us their women, so that we were unable to withstand them: and after having been here that night and half the next day, so great was the number of people who came wondering to behold us that they were beyond counting: and the most aged begged us to go with them to their villages which were further inland, making display of doing us the greatest honour: wherefore we decided to go: and it would be impossible to tell you how much honour they did us.[33]

If such a report were penned in the current century, how might it be received? Anthropologists no longer refer to indigenous peoples as "barbarous" because they have gained insights that were not prevalent in Vespucci's time. Ceremony, song, dance, and what sounded like lamentation, together with rejoicing and great quantities of food, speak of a people whose culture was able to successfully facilitate their pursuit of a quality of life that depended on relationships rather than on mere material acquisition. These songs and dances would not have been just *any* songs and dances but those acquired in ʔuusumč or after some significant event. It must be remembered that, at first contact, the Aboriginals in the New World would have considered these meetings as special spiritual events that had to be appropriately acknowledged with sacred songs and dances. The Nuu-chah-nulth today retain some of these songs and dances, which are displayed only on special occasions. The "lamentations" that accompany these songs and dances are the prayers of petition and of thanksgiving (for the abundance of food), and they provide evidence that these people had discovered a successful way of life – one that had proved itself over millennia.

At yet another place, an island, Vespucci beheld very large footprints in the sand of a beach. As they moved inland, they came upon a small village where there were five females: "two old ones and three girls, so lofty in stature that we gazed at them in astonishment: and when they saw us, so much terror overcame them that they had not even spirit to flee away: and the two old women began to invite us with words, bringing us many things to eat, and they put us in a hut: and they were in stature taller than a tall man."[34] This encounter is most extraordinary from an indigenous perspective. If Vespucci had been looking for signs of an advanced civilization and an intelligent people, he might have been moved by what happened at this time. Typically, the teachings of indigenous peoples emphasize the need to be kind to strangers. In addition, spiritual experiences were commonplace, and the sudden appearance of very strange-looking creatures must indeed have been very frightening, even if not completely unexpected. Moreover, as indicated in the story about the South African shaman Khoisan, who, like the ancient Nuu-chah-nulth, practised responding to fearful situations by facing them rather than by fleeing, the sudden appearance of Vespucci and his party would have had spiritual meaning for the women, although they would not necessarily have understood it at the time. So, according to their teachings, they responded with hospitality and faced up to their fear for, in their system of belief, every unusual personal event had a purpose – even though that purpose might be misinterpreted.

What is said about the possibility of misinterpreting the meaning of events can also be said about the interpretation of scientific observations. Not all scientific reports are valid, and some of the reports that were deemed valid in the past have, over time, proven invalid. There have always been some scientists who are reliable and some who are not, and it is the same with Aboriginal people who profess to be shamans. That some Nuu-chah-nulth shamans are not authentic does not invalidate Nuu-chah-nulth spiritual beliefs and practices any more than inadequate researchers invalidate the enterprise of science. Both, in my view, have proven valid.

To return to the large indigenous women encountered by Vespucci, after their long history of spiritual experiences, how could they have known whether the European explorer and his men were visitors from the spiritual realm or from somewhere else? They acted according to their understanding, just as Vespucci acted according to his understanding, and the two understandings, to this day (with some exceptions), have not yet come together to any significant degree. Such was the impact of Vespucci's letters, read and discussed all over Europe, that misunderstandings were magnified and then swept the globe like a giant literary tsunami, wiping out indigenous languages and peoples from his day down to, and including much of, the twentieth century.

Vespucci's descriptions of Aboriginal life contributed to this misunderstanding because they offered no interpretations from an indigenous perspective. "Their riches consist of birds' plumes of many colors," Vespucci said, assuming that this was the only kind of wealth indigenous peoples owned. Since indigenous peoples were regarded as being so simple, it made sense to conclude that "they neither buy nor sell," which of course was not the case. And, along with the personal freedom that Vespucci observed among the peoples of the New World, the idea that "they [were] liberal in giving" must surely have sparked the imagination of Sir Thomas More when he wrote his *Utopia* – a parable that attacks the chief political and social evils of his time. In fact, as Weatherford points out: "Sir Thomas More incorporated into his 1516 book *Utopia* those characteristics ... in the much-discussed letters of Amerigo Vespucci."[35]

The main character in *Utopia* is Raphael Hythloday, whom More imagined as having been with Amerigo Vespucci on the last of his three voyages to the New World, of which an account had first been printed in 1507, only nine years before the publication of *Utopia*. More writes of Raphael Hythloday: "His patrimony that he was born unto he left to his brethren (for he is Portuguese born), and the desire that he had to see and know the far countries of the world, he joined himself in the company with Amerigo Vespucci."[36] From his description of the Utopians, More leaves no doubt that his inspiration for them was the letters of Vespucci. He uses almost the same phrasing regarding personal freedoms as does Vespucci: "Some of their neighbours ... *live free and under no subjection*."[37] Similarly, certain behaviours and characteristics of the Aboriginals that Vespucci describes are transferred to the Utopians. For example, the Utopians are said to use no trade; to engage in no buying and selling; to be liberal in giving; to have no judicial system; and to offer no punishment to the ill-doer. The behaviours and characteristics noted by More do not exactly parallel those noted in Vespucci's letters, but the resemblance is so close as to render the influence unmistakable. For example, the Utopians do have a judicial system (as, incidentally, did all indigenous peoples), but it was one so simple that each person could be an expert in the law.

Although Sir Thomas More may not have intended to create an exact parallel to indigenous lawmaking in his invented Utopia, he came close to it. Like the legal process in Utopia, the Nuu-chah-nulth legal process was such that each person could be an expert in the law, in the sense that no one could be found to be ignorant of it. However, the Nuu-chah-nulth judicial system did not depend on a centrally located place in which all the laws were kept, as did that of the denizens of Utopia; rather, it depended on the memory of each leader within each big house. These leaders consisted of elderly men and women who were conversant in all the origin stories and in the teachings derived both from them and from that universal method of knowledge, power, and gift acquisition – the *ʔuusumč*.

Of yet another group of Aboriginals Vespucci observes: "They are a people of neat exterior, and clean of body, because of so continually washing themselves as they do."[38] Typically, bathing in a stream, pool, lake, or ocean began as soon as a child was able to run outside. For Nuu-chah-nulth, as stated earlier, this program of early-childhood education was the beginning of training for the more rigorous *ʔuusumč* later in life. In the cooler temperate zones, certain kinds of branches were used like soap to rub the body because they contained tannin, which helped the body to resist the ice-cold waters of mountain streams and pools. When she first married my grandfather, my grandmother went with him to ʔuusumč, and when she emerged from the prayer pool her long black hair became heavy with icicles. In tropical areas and elsewhere, indigenous peoples would use different plants and different rituals for their quests, but all indigenous peoples can be said to hold similar kinds of shamanic beliefs and practices.

In discussing indigenous forms of medicine, Vespucci says simply: "They use in their sicknesses various forms of medicines."[39] Much of traditional medicine was lost to the world because it was thought to be based on superstition. Although it is still practised today, traditional medicine is not nearly as widespread as it was in precontact times.

To conclude, there seems no question about the validity of the theory of *tsawalk,* upon which this chapter is based. Creation is assumed to be a unity in spite of its apparent polarity, in spite of its apparent fragmentation. The theory of *tsawalk* does not propose either to ignore polarity and fragmentation or to manage them through warfare; rather, it proposes to deal with these phenomena by viewing them as essential to life. This view transforms the problem of polarity, the problem of apparent fragmentation, into challenges, which, if met, will result in healthy growth and sustainable living for all creatures. In developmental terms, parallel to the evolutionary concept of improvement, the theory of *tsawalk* assumes that methods used to solve human problems move from those that are immature to those that are mature. The terms "immature" and "mature" are meant to describe the natural development, or evolution, of human societies. It is in this sense of evolution that Genghis Khan's model of a world order is presented. "For the steppe tribes," Weatherford writes, "political, worldly power was inseparable from supernatural power since both sprang from the same source ... In order to find success ... one must first be granted supernatural power from the spirit world."[40] This describes a worldview that, in principle, is identical to that of the ancient Nuu-chah-nulth. Although the conditions of indigenous societies of the past were vastly different from those of contemporary indigenous societies, the principles of freedom and equality described by Vespucci remain the same. As discussed in subsequent chapters, these principles are recognition, consent, continuity, and respect. They are the principles that one can strive to achieve on a daily basis as they represent an ideal standard of human conduct. Of course, human beings are more prone to getting things wrong than to getting things right. This is the human condition (which, it may be remembered, is elegantly told in the Son of Raven origin story). It follows that there is no possibility of attaining a utopia; however, the principles that allow for freedom and equality are worthy goals, even if they cannot be fully realized. The next chapters are an attempt to translate this philosophy into practical constitutional terms.

SHARING RESOURCES
CLEASTOR '11

4

The Nuu-chah-nulth Principle of Recognition

I recall an occasion when my grandmother, Nan, glanced out the window of her home, saw someone coming from a distance, and said, "Here comes Trudy." Since Nan was approaching her seventies at the time, I marvelled at her and said, "I didn't know your eyesight was still so good!" She replied, "I can't see her clearly, but I know how she walks." The phrase, "I know how she walks" provides an excellent metaphor for the subject of this chapter – recognition. The way a person walks in life is critical. Whether that walk is creative or destructive, helpful or disruptive, kind or unkind, depends on the choices that person makes in response to the demands of a seemingly polarized reality. The Nuu-chah-nulth teach that good advice comes over the right shoulder, whereas bad advice comes over the left shoulder. This teaching makes an explicit assumption that reality is inherently polarized and that it requires some form of management. Nan knew Trudy's walk in life.

Whenever Trudy arrived she would always be welcomed by her mother. In saying her daughter's name, Nan would not only acknowledge Trudy's existence but also, as evidenced by her gentle tone of voice, would lovingly accept her. When the greeting was over, Trudy would announce why she had come for a visit, and Nan would pause from her own busy schedule to pay full attention to what her daughter had come to say. If Nan did not completely understand what Trudy was saying, then she would ask for clarification. When Nan understood the communication and Trudy was satisfied with her mother's response, mutual understanding would have been achieved. In this way, good relations were maintained and strengthened. This does not mean that Nan and Trudy did not have disagreements; however, it does mean that they had many points of agreement and that they shared a mutual respect and understanding. In other words, in their relationship we see the mutual recognition of common humanity, not only of its frailty and inevitable fallibility but also of its agreeableness and the opportunities it offers for harmony.

This simple story reveals an important constitutional principle: the importance of recognition, of mutual respect and understanding. The biological relationship of mother and daughter may be a starting point from which to develop a range of relationships, extending from family to friends to nations to international affairs, but it is not sufficient to guarantee a high degree of mutual recognition. What is significant is that any relationship between any two people that is marked by mutual respect and understanding cannot be accomplished without a great deal of work on both sides, whether that relationship is between family members, between friends, or between larger groups.

Formal Recognition

Why is it that gift giving is the core of Nuu-chah-nulth feasts to which visitors from other communities have been invited? And why does the speaker for the host of such feasts, when he calls out the name of one of these visitors, present a gift with words that can be translated into English as: "You have been recognized"? At ceremonial feasts, why is it important to recognize those who have been invited? One answer is that the act of recognition has been found to be an effective way of negotiating a reality that seems to range from utter destructiveness at one end to sublime harmony at the other. The notion that reality is inherently polarized is one that ancient Nuu-chah-nulth accepted as *qua* (that which is). At this level of discussion the focus is not upon the feasting itself but upon the principle of recognition. Feasting is simply one venue within which to enact recognition.

Over time it was learned that gift giving and recognition promoted balance and harmony between beings, that it obeyed what might be called the laws of the positive side of polarity. To neglect the promotion of balance and harmony between beings promoted what might be referred to as the laws of the negative side of polarity. These are not new ideas. Indeed, they are commonly held both by Western and Eastern morality (generosity begets generosity) and by the laws of physics (to every action there is a reaction). When two neighbouring nations shared the same resources, whether cedar, salmon, or human, then it was obvious to the ancient Nuu-chah-nulth that to neglect the act of recognition would open the way to conflict, while to observe the act of recognition, through what I refer to as "mutual concern," would open the way to balance and harmony.

Since the potential for peace and harmony arises in many different social, political, and environmental situations, the Nuu-chah-nulth developed many ways to address issues of relationship. This meant that there were many different kinds of feasts at many different levels of relationship. The wealthiest and most influential chiefs held the formal *international* feasts. In the socio-political arena, these influential chiefs, whose descendants today are known as head chiefs, would formally invite neighbouring chiefs and their families, sometimes years in advance, and always in a personal and ritual way. During these state functions, which might last for several weeks, the business of recognition, which was synonymous with gift giving, became an elaborate and complex institution.

Yet, in addition to the gravity of these formal state functions, the act of recognition and gift giving was also consistently practised in daily living. In principle, there is no distinction between what is practised formally during ceremonial feasts and what is practised informally every day. For example, in ordinary conversation the following remark might be made in Nuu-chah-nulth regarding a social act of recognition: "That person was seen [in a certain location or in the company of friends]." This is a literal translation that includes both the act of looking and the act of realizing that what is observed is a significant relationship. The phrase "that person was seen" is identical to the formal expression of the feast hall, which is translated as "you are recognized." The response to this would be "by whom?" – a question whose purpose is to determine who recognized whom and how this recognition was accomplished. Without exception, these acts of recognition in both formal and informal cases involved biological or legal relationships. So, in this informal case a person may have recognized a distant relative who happened to be passing through the village and instantly offered her/him food and lodging. It would make no difference to this act of recognition whether the relationship was biological or legal as both were regarded as of equal importance. To the ancient Nuu-chah-nulth, because recognition "constituted" a general principle of life, there was a consistent connection between what was formal and what was informal, between what was external and what was internal.

Quu?as and the Environment

The notion of recognition discussed in the preceding section applies equally to forms of recognition between *quu?as* (human) and other life forms. The acts of recognition between Nuu-chah-nulth and other life forms were never perfect but they were often effective. Each season, the salmon were formally welcomed and recognized according to mutually agreed upon protocols. Until very recently, it was assumed that only those beings classified as human were capable of intelligent and meaningful communication. Today, however, a growing body of literature contains scientific findings that appear to corroborate the ancient findings of the *?uusumč*. Jeremy Narby, an anthropologist, writes: "More than a decade ago, I began searching for common ground between indigenous knowledge and Western science, and ended up finding links between shamanism and molecular biology."[1] He was challenged by a biophysicist to test this claim, and so he accompanied three molecular biologists to the Peruvian Amazon, where all "three said the experience of ayahuasca shamanism changed their way of looking at themselves and at the world ... They all received information and advice about their own paths of research ... The two women reported contact with 'plant teachers,' which they experienced as independent entities ... The man said that all things he saw and learned ... were in his mind."[2]

In response to one of Narby's questions about animals, a Peruvian Native said: "But in their world, they are not animals, they too are human beings, and they can speak to one another. They make plans regarding where they are going to go. They check to see if their group is all together, or if one is missing, what happened. Each place where they sleep, they keep lists, they control things as they go. He said that animals and plants contain spirits which humans can see when drinking ayahuasca."[3] Since these investigations into animal and plant intelligence are still in the early stages, there remains considerable scepticism about their findings. What is significant for indigenous peoples is that Peruvian experiences of the world of plants and animals are similar to Nuu-chah-nulth experiences. Both experience plants and animals to have intelligence. Both consider plants and animals to be like people.

In my view, it cannot be overemphasized that these findings from the *ʔuusumč* method of knowledge acquisition form the foundation of indigenous lifeways that have proved sufficiently reliable to sustain ancient communities into the twenty-first century. One of these *ʔuusumč* findings is that all life forms require the development of protocols if balance and harmony are to be achieved. These are the protocols of *tsawalk* (one).

The Origin of Nuu-chah-nulth Recognition

What was obvious from the beginning was the relationship between people connected by marriage and by blood. What was not obvious from the beginning was the relationship between humans and other life forms and, even less so, the intimate relationship between the Creator and the created, between the spiritual domain and the physical domain. The key to understanding the importance of relationships is the interpretation of origin stories, such as the one about Son of Raven and his community, which metaphorically told each listener a parable about how to conduct a *ʔuusumč,* how to acquire the light of understanding, knowledge, power, and the gifts of healing. Through the *ʔuusumč,* it was learned that Salmon, Deer, Wolf, Eagle, Bear, and other life forms are like the Nuu-chah-nulth people. They live.

The story of Son of Raven and his community also taught that resources must be treated in a respectful and lawful manner and that they should not be taken without observing protocols. Underlying all relationships between all life forms is the name Ḱwaaʔuuc, which I translated earlier as a name for the Creator and which literally means "That which is – Owner of All." Ḱwaaʔuuc is the source of all life for Nuu-chah-nulth, and it is through Ḱwaaʔuuc that all life is interrelated. Once this foundation is established, first from direct experiences by members of each household during a *ʔuusumč,* then from public affirmations of these experiences, it becomes both natural and imperative to "see" other life forms in a relational context. In this context, "to see" someone means to observe a significant relationship. A strong Nuu-chah-nulth teaching, therefore, involves always greeting your "blood relatives" with joy and enthusiasm. This ensures that they will always be happy to visit you. It makes sense to treat the salmon in the same way, so that they will continue to return to fulfill a natural and healthy role as a food resource.

Political Science Perspective on Recognition

Theoretically, as understood by political scientists, recognition is, by definition, applicable to state governments.[4] Here recognition is assumed to convey a complex of meanings that begins with the ability of one party to *see* another as important, to *see* value in the other, to consider this value as important, to accept the other on an equitable basis, and thus to be willing to enter into an agreement that can be called a social contract, treaty, or protocol. In 1933, the Montevideo Convention on the Rights and Duties of States stipulated that, to be considered a state, an entity must have a permanent population, a defined territory, a government, and the capacity to enter into relations with other states.[5] This convention has proved a continuing problem for many minority ethnic groups living within a dominant group – that is, the group that controls the government. Hundreds of indigenous groups in the United States and Canada provide examples of this. One recent court case resulted in the Government of Canada making the legal statement: "Nuu-chah-nulth people do not exist."[6] The core of the problem is that recognition, as defined in the Montevideo Convention, is applied only to states and is therefore withheld not only from large numbers of minority groups but also from other life forms, such as plants and animals. The political definition of recognition does exist today; however, from an ancient Nuu-chah-nulth perspective, it has severe limitations.

This chapter proposes that the definition of recognition be extended beyond its current application to statehood to include both peoples and the living environment. While there is no common agreement among scientists regarding whether or not the environment is alive, there seems to be no question regarding its significance to the well-being of life on the planet.

The proposed new definition of recognition is based in Nuu-chah-nulth on-
tology, which has developed over millennia through the testing of myths, or
origin stories, through the *ʔuusumč*. Since an important teaching from the
story of Son of Raven is the need to develop protocols, it is critical to note
that, when Son of Raven is a self-serving egotist, he fails, whereas when
he is clothed in humility, he succeeds. By definition, self-serving people are
blind to the needs of others. They cannot see others because they are fo-
cused on themselves and are unable to *recognize* anyone outside themselves. The Other is excluded. However, a personally developed condition
of humility includes the Other and is therefore a natural process within a
developmental concept that demands struggle, cooperation, hard work, endurance, patience, faith, hope, and vision. Which is one way to describe
life-long learning and education today. The kind of humility that allows
others to exist is natural because others *do* exist and, thus, to deny their
existence is unnatural.

In Son of Raven's case, his humility, expressed in terms of his literal
transformation into a tiny, insignificant leaf – a metaphor for the diminu-
tion of his ego – allowed for the occurrence of mutual recognition. A self-
serving egotism is necessary to babyhood but inappropriate to mature
development. The story of Son of Raven suggests that a natural intention of
creation is to achieve wholeness, to strive for balance and harmony, in the
context of a polarized reality that may have been designed as a necessary
challenge. For example, Son of Raven and his community first addressed
this challenge by striving in a *self-serving* manner to capture the light. After
many attempts using the self-serving method, they finally succeeded with
the "insignificant-leaf," or "humble," method. This process of discovery is
synonymous with healthy growth and development because it unveils the
hidden purpose of life through the lens of humility. Self-serving egotism
provides a lens that magnifies self and so distorts not only the perception of
self but also the perception of others, which is what Son of Raven demon-
strates when he becomes a giant king salmon after everyone had agreed to
transform into the much smaller sockeye salmon. It is the intentional state
of humility, then, that allows for the recognition of others. In this sense,
humility and recognition are indivisible: both are necessary to a natural
process of growth. The implication of this is that contrary forces that are

not adequately managed ensure that mutual recognition cannot happen spontaneously. In fact, the constant challenge, indicated by universal misunderstandings between peoples and nations (seen in history as well as in Nuu-chah-nulth origin stories), is that creation's inherent polarity ensures conflict. Indeed, inherent contrariness is a good definition of conflict and implies that mutual recognition must exist through a dynamic that is capable of balancing and harmonizing opposing forces.

Part of the process of mutual recognition is known as *hamutsup* (unveiling), which ancient Nuu-chah-nulth understood as associated with the need to take *action* in order to realize *intent*. In 1777, Captain Cook did not understand the intent behind the brown skin, dark eyes, strange costumes, rattles, and bird down that he witnessed in Nootka Sound. In this historic *hamutsup* the bird down clearly symbolized peaceful intent. Although Section 35(1) of Canada's Constitution comes some distance towards providing a framework for the recognition of Aboriginal peoples and their rights,[7] there remains the major task of filling in this framework. It is a task of interpretation, of *hamutsup,* on all sides so that Canada and its Aboriginal population can begin to understand one another. In the meantime, although there remains a wall of mutual misunderstanding, of ontological differences, between Canada and its Aboriginal peoples, it is entirely possible to bring down this wall so that each side can begin to see the other more clearly.

Some Obstacles

Some problems are of such long standing, of such complex proportions, and accompanied by such immeasurable pain and suffering that whenever they arise the tendency is to say, with some impatience, "Let's just get past all this!" It was Prime Minister Trudeau who, in 1969, uttered the same kind of sentiments with respect to Aboriginal rights in Canada: "We can't recognize aboriginal rights because no society can be built on historical 'might have beens.'"[8] The prime minister was responding to the outrage expressed by Canada's First Nations leadership to the government's White Paper, which proposed that the terms of settlement for the creation of Canada be forgotten. Trudeau wanted to "get past" all the problems created by the terms of settling Canada, so he suggested that Canada could simply forget its past and start over again as a "Just Society."

The White Paper was unjust because it proposed that the ends justified the means. It was a negative form of recognition that was really a denial of recognition. On the one hand, the Trudeau government's White Paper gave every appearance of offering recognition and equality because it proposed that everyone in Canada should now be treated in the same way; on the other hand, it reeked of non-recognition as there were (and are) still hundreds of outstanding land claims that demand justice. From my perspective, the White Paper appeared to say: "Okay, we've taken most of your land and most of your resources. We've dispossessed you of your languages and cultures. We've submitted you to residential schooling and excluded you from meaningful social, political, and economic participation for most of the twentieth century. And now that you are a beaten, downtrodden people, let's call it even. Let's start over again and we will all be equal." It makes sense, then, that Harold Cardinal, a Cree from Alberta who was a First Nations leader, writer, lawyer, and scholar, countered the White Paper proposal with his book *The Unjust Society: The Tragedy of Canada's Indians*.[9]

Related to the notion of "getting past an unpleasant era" is the idea of ahistoricity. For example, in 1995, a constitutional lawyer presented a lecture on Aboriginal rights during the second year of a new First Nations studies program at Malaspina University-College (now Vancouver Island University) in Nanaimo, British Columbia. The subject under discussion was the landmark case known as *Delgamuukw*.[10] She exclaimed:

I did not understand what they were saying. The elders kept saying, "We want to be recognized! We want to be recognized!" [a pause – then with emphasis] What *were* they talking about? [Another pause as her audience held its collective breath] *Finally!!!* It dawned on me that the "recognition" that they were talking about was the same kind of recognition received at the United Nations by member states and nations. [At this revelation, the tension in the room rushed out the door and was replaced by discernible smiles of satisfaction.] The elders wanted their way of life, their traditional territories, their customs and laws, their histories, their songs and dances, their symbolic regalia, their system of governance, their houses and clans, and their feasting system to be recognized in the same way that other nations are recognized at the UN.

This kind of ahistoricity results, in part, from the deliberate attempt by law, policy, and enforcement to destroy the lifeways of peoples. The constitutional lawyer had difficulty seeing and recognizing the elders as a people who have a history, a legacy of story, song, dance, and what the Nuu-chah-nulth call *haḥuułi,* which refers to sovereignty over bounded territories – the same type of sovereignty that contemporary nations have over their territories. The inability to see and to recognize indigenous peoples is a colonial legacy. In other words, much of contemporary law and politics is based on the hegemonic thinking that was predominant in the Enlightenment and that resulted in the application to Aboriginal peoples of a general state of ahistoricity.

It can be said by indigenous peoples everywhere that, to this point, one of the major obstacles to mutual recognition has been the exclusionary characteristics built into the original constitutions of liberal-democratic states. From the time of the American Declaration of Independence in 1776 to the development of the Canadian Indian Act in 1876, liberal democracy ensured that mutual recognition between Aboriginals and the governments of the United States and Canada would not take place. The Declaration of Independence did this by demonizing the Aboriginal as the problematic Other, while the Indian Act did this by treating the Aboriginal as less than human, or, more specifically, as "other than a person." The evolution of democracy has shown promise, even though it at first excluded everyone but wealthy and privileged males. During colonization, as more European males became wealthy and privileged, democracy broadened its base to include the wealth found in the New World. More recently, under Canada's Charter of Rights and Freedoms, democracy has come to the point of declaring all humans, regardless of gender and ethnicity, to be equal before the law.

Finally, in order to recognize something, in the sense of apprehending and understanding it, it is necessary to have had some prior experience with something similar. Otherwise, recognition is not possible. Captain

Cook's experience with the Nuu-chah-nulth is instructive. Although Cook was able to describe some of the seemingly strange behaviour of the Nuu-chah-nulth, he did not recognize the elements of culture considered vital, such as the type of songs being sung (i.e., prayer songs asking for favour and guidance and also indicating sovereignty over the territorial waters in which Cook found himself). He likewise did not understand that the bird down being spread upon the waters signified intentions of peace rather than of war.

Precedents for Mutual Recognition

In Chapter 1, I noted Thomas Berry's observation that we don't have a good story right now and that this is likely a good reason why the planet is in trouble.[11] A *good* story, I suggest, might be a generic Aboriginal one. If we go back far enough in time, we find that all peoples were at some period Aboriginal – not in the sense of being specifically alike but in the sense that their lifeways are based on mythic story, *ʔuusumč,* songs, dances, chants, medicine, and demonstrable spiritual power. The creative diversity of these expressions is limited only by the capacity of human ingenuity.

A generic indigenous story does not propose to throw out the scientific story but, rather, to complement it. The lifeway of early Nuu-chah-nulth is one example of a generic Aboriginal story, of an alternate interpretation of human origins in which everyone may find precedents for mutual recognition.

Stages of Human Development

Although several characteristics define the early, less mature stages of human development, two of these are pertinent to the development of the capacity for recognition: a lack of experience and a lack of knowledge. Until the first stages of human development are completed, and haven't been interrupted by colonization, a prolonged war, or an extended time of anarchy, a person cannot be expected to engage reciprocally with external reality, to recognize the rights and needs of others, and thus to develop protocols for sharing resources held in common. To gain the necessary experience and knowledge to live together sustainably, in balance and harmony, it may be necessary to have a thorough program of training and education in order to develop each of the interrelated human dimensions of body, soul, and spirit, each of which may have different meanings and interpretations, depending upon worldview and culture. As Elizabet Sahtouris notes,[12] scientists do not agree on what "life" is, so we are left to our own devices and beliefs. But the life that is experienced as soul or spirit is provided some historical context by Carl Jung: "The Latin words animus, spirit, and anima, soul, are the same as the Greek anemos, wind. The other Greek word for wind, pneuma, means also spirit ... There is a similar connection with the Greek psyche, which is related to psycho, to breathe."[13] From a Nuu-chah-nulth perspective, what is important here is not any strict or precise agreement about the definition of soul and spirit but, rather, the experiences of soul and spirit being held in common. As Jung notes in this regard, "I am far from knowing what spirit is in itself."[14] Yet this did not prevent him from developing an entire school of psychological thought, which persists to this day.

Readiness

At contact, European and Aboriginal nations began a relationship that was marked, in very large measure, by a lack of experience and knowledge about one another. This lack of experience and knowledge is also what defines an immature stage of social development. Neither side was ready for the meeting. In the late twentieth century, in an attempt to address such issues of misunderstanding, the United Nations declared that indigenous peoples have the following rights, as articulated in Article 26 of the Report of the Human Rights Council:

1. Indigenous peoples have the right to the lands, territories and resources which they have traditionally owned, occupied or otherwise used or acquired.

2. Indigenous peoples have the right to own, use, develop and control the lands, territories and resources that they possess by reason of

traditional ownership or other traditional occupation or use, as well as those which they have otherwise acquired.

3. States shall give legal recognition and protection to these lands, territories and resources. Such recognition shall be conducted with due respect to the customs, traditions and land tenure systems of the indigenous peoples concerned.[15]

The declaration seems to be saying that indigenous peoples everywhere should become independent, self-sufficient little nations with their own governments, currencies, infrastructures, institutions, languages, cultures, and so on. It is true that, for millennia, indigenous societies throughout the world were self-sufficient, but their noblest struggles were always towards balance and harmony in the context of a sometimes very harsh and always polarized reality. They sometimes failed, some more so than others, and they did not have all the answers to enable them to succeed completely, but they certainly had some of the principles.

One of these principles is recognition. It is to be hoped that, in time, the relationship between indigenous and dominant societies will be marked by experience of and knowledge about each other, which at least would offer the opportunity to develop an ability to recognize the value in each other.

5
The Nuu-chah-nulth
Principle of Consent

"That's just his way!" or "that's just the way she is!" These are possible translations of *qʷaasasa iš!* – a common phrase used to describe characteristic behaviour within traditional Nuu-chah-nulth society. In the context of a worldview that perceives reality to have meaning and purpose and understands everything as happening for a reason, *qʷaasasa iš* takes on multiple meanings. Beyond its ordinary description of characteristic human behaviours, such as those observed between family members on various formal and informal occasions, *qʷaasasa iš* also implies recognition that the purpose of life forms is not always evident or clear to observers. What may be clear are those behaviours that are recognizable within a known value framework; what may not be clear are those behaviours that are not recognizable within a known value framework. Such behaviour is called, in everyday language, acting outside the box.

Used in this way, *qʷaasasa iš* refers to a democratic form of consent defined by a range of behaviour that is mutually agreeable and reciprocal within a society. There is implied in *qʷaasasa iš* an accord, an agreement, a kind of consensus that reality is characterized by purposeful diversity. This is what allowed foreign observers like Amerigo Vespucci to remark that, in such a society, each person appeared to be "lord of himself." At first glance, this notion that each person is allowed to think and do as he or she pleases appears to be anarchic. But this personal freedom is recognized in Nuu-chah-nulth society as limited by the natural laws of creation, such as the law of generosity or the teaching to be friendly, or the admonition to ask for help when help is needed. A major example of a society's adherence to this view of reality is that behaviour that is not understood is still acceptable – at least within "reasonable limits," to use the language of Canada's

Charter of Rights and Freedoms.[1] It is taken for granted that, within the bounds of natural law, there can be an infinite number of possibilities. Consequently, regardless of whether or not a certain behaviour is understood, the traditional response is one of tolerance, of acceptance, albeit within reasonable limits.

While *qʷaasasa iš* acknowledges multiple purposes and roles and allows for the intended diversity of life, its usage here is confined to the issue of consent. In the Nuu-chah-nulth cosmology, *qʷaasasa iš* can be pragmatically applied to behaviour that, by general agreement, is mutually consensual. Consequently, with regard to what is permitted, approved, and recommended, the phrase *qʷaasasa iš* is used in a democratic sense. Not all human behaviour is acceptable.

Tolerance, as indicated by the usage of *qʷaasasa iš,* can have quite a range, from the tolerance for extreme violence exemplified by Canada's practice of cultural genocide against Aboriginal peoples to the tolerance exemplified by indigenous peoples who accepted Canada's formal apology, delivered on 11 June 2008, for its role in the residential school system (which proved so devastating to Aboriginals in so many ways – personally, culturally, socially, economically, politically, and spiritually). Tolerance, in this latter usage, is another dimension of consent and may illustrate a confidence, based in vision-quest findings, that all of life's events, including those that are devastating, may have a hidden purpose.

In the Nuu-chah-nulth story of *Aɬmaquuʔas,* the community of Ahous was destroyed because its children were abducted (in much the same way that children were forced to go to residential schools).[2] In the mythic times of *Aɬmaquuʔas* the community's loss of its children fell within the range of tolerance expressed by *qʷaasasa iš*. The community of Ahous was helpless in the face of the greater power of the giant *Aɬmaquuʔas,* yet its response to her abduction of its children was unquestioning. In such cases of extremity, which can be sudden and inexplicable, it was unthinkable to question the cosmological order. Unlike many of us, who have complained about our bad treatment under the Indian Act, the people of Ahous did not complain that they were being unfairly treated. This is because they assumed that reality is always purposeful.

Today, psychologists understand the condition of complete human help-lessness as a significant factor in human development since it highlights a healthy need for interdependence and interrelatedness, without which human growth may be blocked or stunted.[3] Hence, an important Nuu-chah-nulth teaching, which is often forgotten today, is to ask for help whenever help is needed, not just when things become exceedingly difficult. The abil-ity to ask for help strengthens any family or community, and it is also ne-cessary to personal growth and to developing a measure of independence. Strong individuals who are occasionally in need of assistance make for a strong society. In Nuu-chah-nulth society, being branded as unkind, anti-social, or disconnected, like being branded greedy and ungenerous, is syn-onymous with being suicidal.

*Q*ʷ*aasasa iš* may also be applied to the democratic principle of consent in that it expresses the belief that each individual is unique and that this uniqueness requires free expression. Here, "free expression" is synonymous with "self-expression," but in this case the goal of self-expression cannot be inconsistent with creating balance and harmony within the diversity of community. *Q*ʷ*aasasa iš,* as a common household saying, reminded people of this purpose. Consequently, if it is remembered that the phrase "each individual" refers to all forms of life, to everything that is self-organizing, whether human, plant, or animal, then *q*ʷ*aasasa iš* may be seen to be ap-plicable universally. "Free expression" has precisely the same integrity of being as does a healthy human cell, which, according to Elisabet Sahtouris, makes its own decisions, replicates itself, takes what it needs from its en-vironment, expels what it doesn't need, and exchanges information with its neighbours.[4] Consequently, *q*ʷ*aasasa iš* may also apply to the way everything seems to be self-organizing, the way everything seems to have purpose.

In the days of my childhood there was a man who today might be con-sidered strange and be the subject of public ridicule. As the Nuu-chah-nulth people gradually became infected with the Western perspective on real-ity, some of the children became afraid of this man because his eyes were mostly closed and, in order to see a little bit, he had to tilt his head back. In addition, he had a limp and carried a cane for support, so that as he approached on the road he presented a frightening figure to those not familiar with him. Just one generation previously, this man had been held in universal esteem as a person of great spiritual power. As a little child, I was informed about this man as I often heard my grandparents discuss him with respect. Consequently, when our paths crossed on the road I saw a highly respected rather than a grotesque man. The Nuu-chah-nulth belief system enabled all those who practised it to look beyond gross physical differences. The assumption underlying $q^w aasasa$ $i\check{s}$ (that's just the way she/he is) is the belief that physical and non-physical reality are integrated, and this belief enabled Nuu-chah-nulth to balance outward appearances with inner characteristics. As a result, $q^w aasasa$ $i\check{s}$ allowed for a range of physical diversity within any given community.

In my last visit to the city of Accra, Ghana, in March 2006, I had dinner with one of the Ghanaian professors from the local university. We were accompanied by an American academic who was involved in Third World aid and training. The Ghanaian professor related a personal story of his recent visit to England, where he said that some of the people he met ridiculed him as a living example of a specimen located near the beginning of the Darwinian evolutionary scale. They asked him whether he still lived in trees like his close relative, the monkey. The American academic refused to believe the story since he had just been grandly affirming a very close working relationship with this professor. In fact, he said there was obviously some mistake – the story could not be true. At this denial of his own experience and reality, the Ghanaian professor became quite incensed. He *was* asked if he and his people still lived in trees like the monkeys. There was no mistake about that. Apparently, in the eyes of some English people, this Ghanaian professor, whose colour approaches the purest black, seemed to display significant genetic differences from those common to them. It appears then, in the context of evolutionary theory, that physical difference is not acceptable and, by extension, cannot be a part of consent.

In the context of these stories about a Nuu-chah-nulth and a Ghanaian professor, the meaning of $q^w aasasa$ $i\check{s}$ goes beyond surface appearances to the very nature of being. Darwin's theory of evolution and its interpretations created, for colonizers, a view of differences between people that was and is characterized by superiority and inferiority. A direct consequence of this worldview, supported by nineteenth-century science, was the justification of gross violations of the principle of consent. As suggested by the contemporary experience of the professor in Ghana and Canadian Aboriginal peoples under the Indian Act, the impact of these violations continues to this day.

Qʷaasasa iš is particularly applicable to the academic notion of freedom of thought. This holds even for people with radical views, which are measured against community standards. For example, during the 1950s, a man stood up in an Ahousaht council meeting and declared: "We should separate from Canada and seek foreign aid from China." There was shocked silence – then wonderment. In a society based on *ʔuusumč*, people wondered if the man with the radical views might have had access to information not currently available to others. It was best to wonder and wait for further developments. What is interesting here is that a new and untested idea was tempered by ancient ideas that had been tested over long periods of time and so had formed a framework for sustainable living. One of these ancient ideas is found in *ʔuu*, which is the root of the word *ʔuusumč* (vision quest) and which means "to be careful." A common expression, *ƛuuɫhapii*, meaning "go slowly," reflects this idea.

In a governance context, *qʷaasasa iš* can be understood to mean a form of democratic consent that allows, by community agreement, freedom of movement, freedom of speech, freedom of thought, freedom to be unique, freedom to grow, and all those freedoms-of-being that do not violate the freedom of others to be. In fact, it was more than three decades later, after this man had passed into the other world, that Ahousaht finally began to move away from colonial rule towards some semblance of self-determination. Ahousaht did not separate from Canada or ask for foreign aid from China; rather, it moved slowly, according to its ancient tenets.

Qʷaasasa iš can also be translated as "freedom to believe." Carobeth Laird's *Mirror and Pattern: George Laird's World of Chemehuevi Mythology* makes a comparison between Chemehuevi belief and Christian belief: "The myths, therefore, were the pattern and are for us the mirror of a culture which has perished from the earth ... They were the unwritten Bible by which the people must live."[5] In Chemehuevi mythology, there were three primal brothers: Wolf, Mountain Lion, and Coyote. Wolf and

Mountain Lion display the admirable qualities – wisdom, courage, patience, restraint, insight, power, strength, and humility – whereas Coyote represents the self-absorbed, egotistical, greedy, grasping, scheming personage always out for an easy meal, a shortcut, or a quick solution to any and every problem. Although these contrasting characteristics-of-being appear grossly stereotypical today, traditional indigenous peoples accepted them as standards, as accurate reflections of important elements of reality. They are *qʷaasasa aɬ,* just part of the way people are. The Chemehuevi declare that, after the mythic age passed, they followed Coyote rather than Wolf.

The dichotomy between the character of Coyote and the characters of Wolf and Mountain Lion, which are reflected in the nature of being, necessitates that this dichotomy be managed through conscious choices. This natural tension in the nature of being – between good and evil, creation and destruction – demands choosing between consent or coercion, consent or oppression, consent or dictatorship. The Wolf and Mountain Lion archetypes represent consent, whereas the Coyote archetype represents a violation of consent.

The Nuu-chah-nulth story of Son of Raven and his community contains the same mythical powers as do the Chemehuevi stories about Wolf, Mountain Lion, and Coyote. In the Chemehuevi stories, each main character only needs to think in order to execute an action. Blue Bird thinks that Coyote should look upward, and immediately Coyote looks skyward. Whatever Wolf thinks of someone else as doing is what is immediately enacted by that person. In the Nuu-chah-nulth story of Son of Raven and his community, when it is decided that everyone should turn into sockeye salmon, there is no explanation as to how this is done. There is no discussion about physical impossibilities or miracles or magic. The mere thought of everyone's transforming into sockeye salmon is sufficient to complete the act. These acts of transformation represent a perfect balance between the physical and non-physical aspects of creation.

Son of Raven is successful only when he takes the advice of Wren, whose name means "One Who Always Speaks Rightly," which is to become as small a being as he can be – an insignificant leaf floating in a spring well. Humbling oneself is a typical Wolf-like choice, and this act of humility on the part of Son of Raven stands in sharp contrast to his previous act of aggrandizement, whereby he presumed, contrary to the original agreement, to transform into a giant king salmon rather than into the smaller sockeye salmon.

The belief about which road in life leads to destruction and which road leads to creation is graphically told in the Nuu-chah-nulth story about Pitch Woman (*Aɬmaquuʔas*) and Son of Mucus (*ʕintḥtinm'it*).[6] Pitch Woman lived in the mountains and periodically came down to steal children while they were at play on the beaches. Apparently she was a giant, for she carried a large basket on her back into which she threw the children before carting them up the mountain to her home. Pitch Woman was an archetype of evil, taking children against their will. In due time, Son of Mucus was miraculously born in a mussel shell and became the son of the local chief. When he grew into young manhood, Pitch Woman came again and stole more children. This was the reason that Son of Mucus had been sent from above to be born into a chief's house. After he killed Pitch Woman and brought the children of Ahous back home, he enlisted the aid of the villagers to make a very large number of arrows. With these arrows he ascended into the sky, returning to his original home, and in so doing showed the preferred way of life. Like Son of Raven when he submitted himself to the household of the Wolf chief, Son of Mucus did not intervene in the affairs of the community at Ahous without its consent. As a great spirit-being, Son of Mucus had the power to force himself into any situation, but if he had chosen that route he would have violated an important principle of Nuu-chah-nulth life – the principle of consent. So Son of Mucus chose to be born into the Ahous community and to submit himself to its way of life; learn its language, customs, and teachings; and become a vital part of the Ahous family. In this way, when the time came to rescue the children, Son of Mucus was able to do so as a legitimate member of the community. He had legal and familial rights that gave him the necessary communal consent to restore the children to their families and community.

Qʷaasasa iš also describes a form of consent between animals. In a Nuu-chah-nulth origin story wolves are chasing Deer, who escapes into a tree. When the tree is cut down by the wolves, Deer jumps into another tree, and this process continues until the wolves are advised by Wren to sing a song, which brings Deer crashing to the ground. At this point – for a natural order is assumed in these stories – consent is granted to Deer's request that, in exchange for giving up his body, the wolves would leave his stomach intact. This agreement, like the creation of a natural constitution that is co-scripted by two separate beings, has been, until recently, honoured by both wolves and Nuu-chah-nulth hunters. On the surface, this does not appear to be true consent but, rather, coercive bullying; however, beyond the surface, beyond physical reality and the apparent destructiveness of the creative process, there is a wholeness, a unity, that balances and harmonizes what appears (on the surface) to be conflict.

On a recent occasion, my Aunt Trudy's husband, Edwin, and I were discussing an issue of imbalance that had been created on the coast of British Columbia by the introduction of sea otters, which had at one time become very scarce but were now overpopulating the coastline and devouring most of its sea urchins.[7] When the population of a species explodes out of control, biologists describe it as "invasive." In the course of our conversation, Edwin told me a short version of a story that could be about sustainability through the development of protocols between consenting parties.

An Origin Story about Bear: Sustainability, Sharing, and Protocols

A man owned a fish trap placed in a river. The trap was designed to capture enough fish to meet the needs of the man's extended family rather than in the contemporary manner of fish traps, which are designed for maximum exploitation for maximum profit. During the salmon run season the man checked his trap each day only to find that it had been emptied of fish and wrecked in the process. This continued for some time until finally the man decided to stay overnight by the fish trap to see what was happening.

Sure enough, during the night the man saw Bear go to his trap and take the captured fish in such a way as to wreck the trap in the process. The man approached Bear and asked why he was taking the fish from his traps. Rather than explain in words Bear suggested that the man accompany him so that he could show why he was taking the fish.

After trekking for some time the man grew tired so Bear put him on his shoulders and they travelled a great distance, over four mountains, until they

came to a village of many people. All of these people were like Bear. They wore
bear skins whenever they went somewhere and took their skins off when they got
home. The fish from the trap was given to provide for this village. The man stayed
and observed all the activity, the way that the fish were prepared and served and
the good use that was made of them.

After some time, Bear said that he would take the man back but that first
the man must meet the chief of the village. At this meeting an agreement was
struck between the man and the bear nation. If the man was willing to share
from the bounty of his fish trap, there would always be plenty of fish for both
tribes, the man's tribe and Bear's tribe. When Bear returned the man to his river,
sure enough, the fish trap was full and half of it went to Bear for his tribe. Part
of the agreement was that Bear would no longer wreck the trap so that equitable
sharing might continue in perpetuity.

This story of Bear and the fish trap has some of the same characteristics as
the story of Son of Raven and his community. In the beginning, Bear does
not seem to know how to negotiate the reality in which he finds himself.
Bear is out looking for fish and discovers the fish trap and helps himself
to the resources that he needs. Every time Bear returns to the scene where
he first found the fish trap, it has been repaired, and, as a result, more fish
are to be had. In the story of Son of Raven and his community, all know
where the light is kept, and, like Bear, they begin to attempt to take it without any regard for its owner. In the course of events both Bear and Son of
Raven learn about the natural need to obtain consent by developing protocols with their neighbours.

Another example of consent between humans and animals is the relationship between Nuu-chah-nulth whaler and whale. In *Tsawalk*, there is an
account of Keesta, my great-grandfather, who, after eight months of
ʔuusumč, struck an agreement with the spirit of a whale. In exchange for
being recognized as a great personage, for having great honour bestowed
upon it, and for undergoing the transformation natural to creation, the
whale agreed to be captured. That is why, when a whaler like Keesta throws
the harpoon, it is said that the whale "grabs" it and "holds" on to it. This is
because the harpoon entering the whale is the practical fulfilment of an

agreement and is a good example of consent. The whale consents to this agreement. Although the agreement is stated simply here, the enterprise is fraught with danger and risks. For example, it was necessary for all eight paddlers, along with the whaling chief, to be cleansed, purified, and in the right spirit. If one member of the team happened to be out of spiritual synchrony with the rest, this could jeopardize the entire venture and possibly result in loss of life.

In the contemporary context of environmental crisis highlighted by climate change, consent between humans and other life forms, which is common to the ancient Nuu-chah-nulth, may be considered a working principle of sustainability. For example, consider what occurs when Nuu-chah-nulth take down a great cedar tree to transform it into a canoe or into parts of a house and household items and utensils. The tree is acknowledged as a personage and paid much respect and honour. Although the tree is essentially a captive resource that has no power of resistance, it is nevertheless recognized as an *organic and living* part of a larger entity that scientists today refer to as nature. Because a tree is a member of a larger entity, what one does to it is also done to nature, in the same way that anything done to a human hand, foot, ear, or eye is also done to the entire human body. When a tree is cut down without recognition of, and respect for, its life force, nature is negatively affected.

Many scientists view nature as an organic entity created through an evolutionary process. In this worldview humans and other life forms may coexist in physical space but are necessarily separated by an evolutionary scale. A traditional Nuu-chah-nulth perspective on nature is that it is a *living* entity. As a result, in the words of Roy Haiyupis, a deceased Nuu-chah-nulth elder, nature, when abused, will "strike back." Nature lives and will respond in kind to the way it is treated. Climate change, pollution, violent storms, floods, earthquakes, dying rivers, lakes, and streams are signs of nature's response to abuse suffered through violations of the principle of consent.

The Western Principle of Consent

Consent is central to the democratic tradition and is both a legal right and a privilege. Consent stands in direct opposition to arbitrary rule and slavery.

Where arbitrary rule ignores the opinions and voices of a general populace, consent demands that those in power consider these same opinions and voices in ways that translate into political goods. In practice this means that, everything being equal, consent is an unquestionable constitutional ideal with regard to relationships between people and nations. Unfortunately, everything is not equal, in spite of the declaration of equality in Section 15 of Canada's Charter of Rights and Freedoms. Constitutional ideals have not been easy to translate into practice. One of the most likely reasons for this difficulty is that the universal application of the democratic ideal of consent is a recent political phenomenon. Most of the previous five hundred years of colonization can be characterized as a period of continuous violation of consent towards women in general and Aboriginals and other dispossessed groups in particular.

Western democracies have had little practical experience with the implementation of democratic ideals that apply not only to every person but also to every group. There has been plenty of practice in implementing democratic ideals within racial boundaries. Englishmen in the early days of Canada limited the application of democratic ideals to themselves. In fact, based on their immigration policies, there is reason to argue that, during the nineteenth century and for much of the twentieth century, the Western democracies, which included Britain, France, the United States, and Canada, were more concerned with maintaining their racial boundaries than they were with universal democratic ideals. When "others" were allowed into these countries, it was so that they could be exploited as cheap labour to serve the ruling classes. It was for this reason that blacks were imported as slaves into the United States, that the Chinese were allowed to enter Canada to work on the railways. During the nineteenth century and the early days of settlement in British Columbia, much of the labour force was made up of First Nations people because there were not enough settlers to fill the necessary jobs. As soon as the settler population increased so that it could take care of labour requirements, First Nations people were no longer necessary and, for the most part, joined the forces of the unemployed.

As indicated in the previous chapter, a major problem with the practice of democracy is that of recognition. When a people are not recognized, it is difficult, if not impossible, to allow them the privilege of consent. The necessary relationship between recognition and consent may not be immediately evident because we now live in a world in which mutual recognition between states is commonplace on the world stage. If it is common practice for one state to recognize another in spite of language and cultural differences, why should there be any necessary relationship between recognition and consent? Why cannot one state recognize another without the provision of consent? Yes, it is possible for one state to recognize another without the provision of consent, but the Nuu-chah-nulth point is that recognition is a precondition of consent. Ancient Nuu-chah-nulth peoples did not at first recognize any relationship with the Wolf community, but, after some fundamental experiences during *?uusumč*, they learned about their relationships with these others and were then able to properly recognize them. Prior to this learning process, the ancient peoples tried to take what they needed without consent. This was true both for Son of Raven and for Bear. Even in the material world of power politics, one may ask if women could have achieved suffrage without first being recognized by their male counterparts. Not likely.

The Aboriginal peoples of Canada are another case in point. The federal franchise was not granted to indigenous peoples until 1962, under the Conservative government, and could not be exercised until the elections of 1964. The right to vote is central to the democratic principle of consent. There can be no democracy without consent. Yet, even though the Aboriginal peoples of Canada received the federal franchise in 1962, they were still bound by the constitutional terms and conditions of the Indian Act, 1876, which was a violation of the spirit of the original treaties with the British Crown and the democratic principle of consent. According to the Royal Proclamation of 1763, one central intent of any given treaty was "that the several Nations or Tribes of Indians with whom We are connected, and who live under our Protection, should not be molested or disturbed in the Possession of such Parts of Our Dominions and Territories as, not having been ceded to or purchased by Us, are reserved to them, or any of them, as their Hunting Grounds."[8]

Accordingly, the understanding of many First Nations peoples was that their ancient lifeways would not be disturbed. That intent seems quite clear, but now, with historical hindsight, we learn that often First Nations peoples' understanding of treaties was not shared by European powers. Nevertheless, when a group is not to be disturbed, it means that there is to be no intervention in or deliberate disruption of its lifeways. To the Nuu-chah-nulth, if they hear that a group is not to be disturbed, then they assume that that group is to be left alone so that its members can continue their ancient cultural practices in their own way. It does not appear to me possible that there could be any misinterpretation or misunderstanding of the meaning of "should not be molested or disturbed." Yet, misinterpretation or misunderstanding regarding the intent of the treaties, a process controlled completely by the dominant party, continues to be the order of the day and continues to divide non-Aboriginal and Aboriginal peoples without any apparent solution in sight.

Although what follows is all too familiar to some, it unfortunately remains an authentic and substantive part of what is called Canada today. It is presented in the spirit of the *λuuk*ʷ*aana,* as in: "We remember reality." In 1876, the Indian Act attempted to consolidate various statutes concerning Indians and officially made the First Nations of Canada into wards of the state. The Indian Act is a transgression of the treaties made in good faith between the British Crown and many of the indigenous peoples of Canada. What came into being in 1876 was clearly a case of molestation, whereby the lifeways of the First Nations in Canada were "disturbed." From a First Nations perspective, there could have been no greater intervention into indigenous lifeways than the banning of the potlatch, as stipulated by an 1884 amendment of the Indian Act, 1880:

> 3. Every Indian or other person who engages in or assists in celebrating the Indian festival known as the "potlach" or in the Indian dance known as the "Tamanawas" is guilty of a misdemeanor, and shall be liable to imprisonment for a gaol or other place of confinement; and any Indian or other person who encourages, either directly or indirectly, an Indian or Indians to get up such a festival or dance, or to celebrate the same, or who shall assist in the celebration of the same is guilty of an offence, and shall be liable to the same punishment.[9]

Sir John A. Macdonald's comments about the potlatch during a House of Commons debate on 19 April of 1884 expressed the prevailing view of the day: "The third clause [of Bill 87] provides that celebrating the 'Potlatch' is a misdemeanor. This Indian festival is a debauchery of the worst kind, and the departmental officers and all the clergymen unite in affirming that it is absolutely necessary to put this practise down."[10]

Even now, although more than half a century has elapsed since the repeal of this law in 1951, the potlatch ban is difficult for me to discuss because it had such traumatic consequences for my family in particular and for First Nations in general. While for some scholars the issues are academic and without personal meaning, there is likely no effective way to convey the full meaning of this ban except to compare it to an unthinkable act that would have similarly disastrous consequences – a ban of both the institution and practice of the Canadian Parliament. In addition, imagine what might happen if Canadians were directed by policy to learn a new language and to forget the language they had always spoken.

In the early 1940s, during the potlatch ban, a chief's family from a neighbouring community to the north came to *lučhaa* – that is, to ask for the hand of my aunt in marriage. Under ordinary Nuu-chah-nulth cultural circumstances and conditions, this act of *lučhaa* would not be remarkable except to those directly involved. To observers of the day, the *lučhaa* was conducted in the accepted and proper manner according to known protocols. Years before, at some potlatch or other ceremonial gathering, the parents may have discussed marriage arrangements. People who came to *lučhaa* were always anticipated, just as Son of Mucus's *lučhaa* was expected by the chief in the spiritual realm.[11] Marriage was never left to chance or to immature, inexperienced youth. In the present case, the people who came to camp around our house to *lučhaa* stayed the customary four days, singing their songs and making speeches about the suitability of their son for my grandfather's daughter.

However, for my aunt to be married in our cultural way, it would have been necessary to have a potlatch, then a criminal offence. A complication to this otherwise normal event was that my aunt was fourteen years old, of marriageable age in our culture but too young to be out of school in Western culture. According to Section 9, subsection 2 of the Indian Act:

2. Such regulations, in addition to any other provisions deemed expedient, may provide for the arrest and conveyance to school, and detention there, of truant children and of children who are prevented by their parents or guardians from attending; and such regulations may provide for the punishment, upon summary conviction, by fine or imprisonment, or both, of parents and guardians, or persons having the charge of children, who fail, refuse, or neglect to cause such children to attend school.[12]

Parents could be sent to jail if their children did not attend school. In this case, my grandfather could also become a criminal if he chose to follow through with the marriage of his daughter by putting on a potlatch. My aunt never did marry. My grandfather died a few years later in an alcohol-related incident, the sort of thing that continues to be a common experience under the governance system imposed by the Indian Act.

Due to the powers of the Indian Act, which were backed by the powers of the Canadian state, enforced by the Royal Canadian Mounted Police, and administered by Indian agents, the powers of people like my grandfather soon declined, and social chaos gradually encroached upon our community like a tsunami sweeping away millennia of accrued wisdom. Even though my grandfather and his fellow chiefs initially adapted the Indian Act's imposed band council system so that it would fit with their customary form of governance,[13] the Indian Act's system eventually prevailed and replaced the more effective form of governance familiar to my grandfather. When the management system of a culture is criminalized, social chaos is the outcome. The greater technological powers of the colonizers served to humiliate the lesser powers of Aboriginal chiefs. Every Nuu-chah-nulth chief's reputation as a leader was dependent on an ability to provide for the well-being of his or her people. Consequently, the more the Indian Act usurped local powers, the greater the humiliation of the Nuu-chah-nulth chiefs and the lower they fell in the esteem of their own people.

Since local chiefs were constitutionally disempowered by Canada through the Indian Act, and since the actual constituted authority over the lives of First Nations people was not present on the reserves, Nuu-chah-nulth teachings, practices, and belief systems began to erode. This erosion took place in spite of the fact that Brian E. Titley, professor of education at the University of Lethbridge, calls the Indian Act's prohibition of the pot-latch a meaningless piece of legislation. According to Titley: "Indian agents refused to seek prosecutions because they believed that the forbidden cus-toms were harmless. Others feared that repressive measures would precipi-tate an Indian uprising. The minimal police presence on the coast, the reluctance of the provincial authorities to co-operate, and the absence of jails and guard-houses were additional factors that rendered the legislation meaningless."[14] Titley's point has scholarly merit, but I cannot agree that the potlatch ban was "meaningless." I lived under this ban with my ex-tended family and witnessed its destructive powers as I moved from the stability of my early years to the terrifying instability of a community whose laws had been displaced. The potlatch ban helped to erode the great teachings of Nuu-chah-nulth, which were based on *iisʔak̓* – respect for all life forms.

At the personal level, where we all must live, the most immediate effect of the erosion of the value *iisʔak̓* for the Nuu-chah-nulth was a correspond-ing erosion of the ancient practice of being kind to friends, neighbours, strangers, and relatives; of honouring the young female who was con-sidered *usma* (precious); of honouring elders and grandparents; and of al-ways loving your relatives. The social chaos that occurred on reserves over a century under the auspices of the Indian Act, especially after the potlatch ban, is reminiscent of the social chaos evident in Baghdad as a direct result of the downfall of the Iraqi government in April 2003. Hereditary chiefs and their families could be disregarded and openly ridiculed, as I often was, being an heir apparent to a chief's seat. The wise counsel of elders and grandparents went unheeded, and they, too, were often ridiculed because they were thought to be examples of beings living at the wrong end of the evolutionary scale.

Widespread alcohol abuse, although it could temporarily reinforce cul-tural lifeways through a false sense of well-being, a false sense of adher-ence to *ʔaapḥii* (the teaching that we are to be kind to others), would lead to a gradual disintegration of the once strong cultural fabric supported by the institution of the potlatch. Over time, children learned to live in fear where they once felt safe, secure, and nurtured. Women could be abused at will if they had no male protector. The elderly became unable to fulfill their role as advisers, and, on isolated reserves (of which there are many in Canada), in the absence of on-site authority the ancient rule of law gave way to the arbitrary rule of force and coercion (a phenomenon that may be observed in unsupervised schoolyards during recess). In their conduct, First Nations people came more and more to resemble the imaginary descriptions of Aboriginals posited by Rousseau.

Under these kinds of social conditions, the rule of law, so far removed in Ottawa and in other urban locations in Canada, no longer applied. If one has always enjoyed the rule of law, it may not be possible to imagine the horror of its absence. I have experience of its absence both on my reserve and in residential school. Since I am an eldest son in the house of *ƛaaqišpiiƚ*, I was born into a hereditary chief's position. As a teenager, during summer holidays on the reserve, this accident of birth sometimes became a source of torment from older boys who did not enjoy the same kind of traditional position. The suggestion is not that this kind of negative attitude was absent in the ancient past, for great good and great evil co-exist within the cultural fabric of Nuu-chah-nulth society, but that during the 1950s in particular, the administration of the Indian Act so eroded the ancient governance system that young people were encouraged to disregard its authority.

Similarly, in residential schools, which were set up without the consent of First Nations parents, whenever supervisors were absent, little ones were always at the complete mercy of older boys. Bullying, intimidation, beatings, and sexual abuse were common. During the 1940s, my grandfather first lost his eldest son to a violent storm and then his eldest grandson, myself, to a residential school. No First Nations parent or community, either prior to contact, during contact, or after contact, has ever disagreed with the principle of education. It has always been considered a necessary part of life. Yet the questions remain: How is it that a whole continent of Aboriginal peoples could allow their children to be educated in foreign residential schools? How could they allow their children to go to these schools under the condition that they no longer speak their own languages and that they learn a new language?

One answer to these questions is that, from the outset, it appears that Aboriginals in general were impressed with the superior technology of the newcomers. The new steel and iron tools and implements were immediately adopted because it was practical to do so. The new technologies improved the quality of life. To the Nuu-chah-nulth, the first ships that came to the west coast of Vancouver Island were marvels of technology. The word *Mamaɫn'i,* used to describe white people, can be translated as "people of houseboat," or simply as "boat people." Prior to contact, the Nuu-chah-nulth had no experience of watercraft containing cabins large enough for human habitation.

The problem for the Nuu-chah-nulth in particular, and likely for Aboriginals in general, was that every demonstration of achievement, great or small, was taken as a sure sign of spiritual power. Thus, when my great-grandfather Keesta brought home a large whale to feed his people, in their eyes this feat was synonymous with spiritual power. Even down to a recent generation of Nuu-chah-nulth, it was common for a successful fisher to remark to another who had not been so fortunate that he must be *wiiša?* (spiritually unprepared). This term is specifically applicable to one who has broken a taboo by having had sexual intercourse prior to the fishing expedition. It is no exaggeration to say, as others have said about Aboriginals in general, that the Nuu-chah-nulth people were (and many still are) highly spiritual in their orientation towards life.[15]

Consequently, when the great European ships sailed into the sovereign waters of the Nuu-chah-nulth territories, the people in these ships *were assumed to have access to spiritual powers.* Although this is a fair explanation of Nuu-chah-nulth thought, the actual historical case was not so clear-cut. From a Nuu-chah-nulth perspective, spiritual power was a mixed bag.

Philip Drucker, an American anthropologist who specialized in Native American peoples, makes the following remarks about the introduction of European technology into the coastal life of the Nuu-chah-nulth during the fur trading era:

> They purveyed to the natives great stores of wealth in the form of metal tools, firearms, and ornaments. They alternately cajoled, robbed, and murdered them. Hannah exchanged names with an Ahousat chief, Meares had his sailmaker rig a suit of sails for "Maquinna's" canoes, Kendrick flattered the Indians by dressing in native garb. The Spaniard fattened

Maquinna's vanity and usurped the site of Yuquot for their garrison. Martinez shot the high-rank chief "Quelquelem" to death for a fancied slight, and Boit, on Gray's orders, shot up and burned out the Clayoquot village. Then traders of every nationality were horrified by the Indian massacres of the ships *Boston* and *Tonquin*.[16]

The advanced technology was useful; the guns, knives, household items, and metal tools were all good. But some of the newcomers were not good people, just as some Nuu-chah-nulth were not good people. For example, as indicated by Drucker, some of the European visitors abused and killed Aboriginal people during the fur trade. It is for this reason, according to Nuu-chah-nulth, that Ahousaht warriors captured and burned the *Kingfisher*, a fur trading ship, in retaliation for abuses committed by fur traders on a previous visit. The English sent a warship to demand that Ahousaht give up the perpetrators of this just act, but the Ahousaht war chief refused. Gilbert Sproat records the incident as follows:

> I give here an extract from a dispatch of Rear-Admiral Denman, dated on board the frigate *Sutlej,* in Clayoquot Sound, the 11th October 1864, and addressed to the Governor [Kennedy] of Vancouver Island. A trading schooner, the *Kingfisher,* had been destroyed [in August] near the shore by Cap-chah, an Ahousaht chief, who told the captain that he had a quantity of oil to dispose of. Cap-chah killed the captain, and the two other Ahousaht killed the sailor who was on board, as well as a Kwakiutl, who was one of the crew.[17]

Sproat records several pages from the official dispatch from the British warships. The superior technology of the British proves destructive to Nuu-chah-nulth homes and canoes, which are destroyed by heavy fire. However, Cap-chah is not captured, nor will the Ahousaht give in to the threats and demands of the British. Nevertheless, an official warship dispatch reads as follows:

> The success of this affair is due to the excellent conduct of Lieutenant Stewart, and the officers and seamen and marines under his command,

while the defeat of the Ahousahts by an attack after their own fashion, has produced profound alarm and astonishment ... I have promised that no further measures shall be adopted against the Ahousahts for one month from this time; but if the six murderers are not given up by that time, I shall be obliged to order forcible measures to be resumed.[18]

If there had been independent observers of this incident, Rear-Admiral Denman's report may not have been so effusive regarding British success. In the end, the warships did not return as promised, and Cap-chah's reputation was greatly enhanced. Ahousaht oral history suggests that, in spite of the superior firepower of the British, in close combat Ahousaht warriors were often more than their equal, which is not surprising, given that they fought on terrain that was familiar to them but unfamiliar to the British. Roy Haiyupis, an Ahousaht elder, now deceased, relates that a skirmish that took place at Herbert Inlet resulted in a retreat by the British.

Over time, oral history was mostly forgotten in the face of constant pressure to erase it through being forcibly educated into new and alien ways of being. Particularly during the post–Second World War era of the 1950s, this constant pressure reduced many Nuu-chah-nulth to feeling a sense of shame for being indigenous, and this may represent our darkest period, the time during which our cultural loss was greatest. Then came the era of protests in North America – Black Power and Red Power and Women's Power – during the 1960s, and it is during this time that some Nuu-chah-nulth began to revive their ancient ways of being, began, in the face of fierce opposition, a process of remembrance. In hindsight it is now apparent that there have always been those who continued to practise Nuu-chah-nulth ways of being for, to this day, the ʔuusumč continues in secret and Nuu-chah-nulth people still testify among themselves to its efficacy.

Still, one conclusion appears inescapable to the Nuu-chah-nulth: those of European descent appear, through their superior technology, to manifest spiritual power. This explains some of the apparent ease with which government and church authorities enforced their residential school policy. What is problematic about this is that the democratic right of First Nations consent was completely absent. First Nations leaders, parents, and communities had no say in, no input into, and often no knowledge about a piece of legislation that would govern every aspect of their lives from birth to death for most of the twentieth century.

Another problem associated with the enforcement of residential school policy was an inappropriate means to a desired end. The education policy found in the Indian Act stipulates that First Nations children should be educated – an end with which no one would disagree. However, the means and type of education were coercive. The children's language, values, culture, and lifeways were all denied, which was a sure way, according to contemporary educational research, to guarantee failure. From the sixteenth century to well into the twentieth century, First Nations children failed massively under this kind of practice. In 1688, Mother de l'Incarnation made the following observations:

> It is however a very difficult thing, although not impossible, to francize or civilize them. We have had more experience in this than any others, and we have remarked that out of a hundred that have passed through our hands scarcely have we civilized one. We find docility and intelligence in them, but when we are least expecting it they climb over our enclosure and go to run the woods with their relatives, where they find the more pleasure than in all the amenities of our French houses. Savage nature is made that way; they cannot be constrained, and if they are they become melancholy and their melancholy makes them sick. Besides, the Savages love their children extra-ordinarily.[19]

About three hundred years later, an updated account of First Nations education was provided in the Hawthorn Report of 1966-67. In this report the national failure rate of First Nations students in the grades 1 to 12 program in Canada was 94 percent. In British Columbia the failure rate was 96 percent. During the 1990s, the Royal Commission on Aboriginal Peoples (RCAP) sought to explain these failure rates in terms of the problems created by Canada's colonial treatment of its Aboriginal peoples. The RCAP recommended four principles as the basis for a renewed relationship between Aboriginals and Canada: recognition, respect, sharing, and responsibility. Except for the peculiarly Aboriginal principle of respect, the other three parallel contemporary constitutional conventions identified by James Tully as recognition, continuity, and consent.[20] The principle of *sharing* found in the RCAP report parallels the constitutional convention of *continuity* because both allow for different communities not only to express their own opinions but also to gain access to common resources. The principle of *responsibility* found in the RCAP report not only parallels the constitutional convention of *consent* but also implies *recognition.*

Since these are apparently sound constitutional principles upon which to base a renewed relationship between First Nations and Canada, one might expect a government to welcome these suggestions. However, to date, although some progress has been made in Canada–First Nations relationships, the Indian Act remains in place. And this is experienced by many who live on reserves as negative.

An issue related to self-government is treaty making, particularly in British Columbia. During the 1990s the government of the Province of British Columbia refused to recognize that First Nations peoples lived in organized societies prior to contact and settlement. In *Delgamuukw v. British Columbia,* Chief Justice Alan McEachern found against the Gitksan-Wet'suwet'en, who were making a claim for some 58,000 square kilometres of traditional land. In March of 1990, he ruled that Aboriginal rights do not include ownership of land and, moreover, that these rights exist at the pleasure of the Crown. Therefore, McEachern ruled, Aboriginal rights are extinguishable "whenever the intention of the Crown to do so is clear and plain."[21]

In 1997, the Supreme Court of Canada overturned McEachern's ruling and affirmed that Aboriginal rights are sui generis and are not dependent upon the Crown for their existence. In fact, just as the Canadian Parliament's justification and existence are founded on British common law, the rationale for the existence of Aboriginal rights is founded on ancient lifeways and on laws derived from Aboriginal teachings, stories, beliefs, and customs. In this respect, any attempt to extinguish Aboriginal rights would be illegal.

Recognition of Aboriginal rights, however, is not always something that can be clearly interpreted. One historical reason for this is that ideas about land ownership differ from one country to another, depending upon culture. Peoples of European descent hold to one set of viewpoints and Aboriginal peoples hold to another. John Locke provided one idea of property rights, which was based on working the land. In 1689, Locke said: "Thus, *Labour*, in the Beginning, *gave a Right of Property*."[22] By "labour" Locke meant working the land for farming purposes. In this view, any land that was not worked was deemed to "lie in common": "God gave the World to Men in Common; but since he gave it them for their benefit, and the greatest Conveniences of Life they were capable to draw from it, it cannot be supposed he meant it should always remain common and uncultivated. He gave it to the use of the Industrious and Rational (and *Labour* was to be *his Title* to it) not to the Fancy or Covetousness of the Quarrelsom and Contentious."[23]

There are two principal ideas about land and land ownership, or title: (1) the working of land leads to entitlement to that land; and (2) land that is not worked is not owned by anyone – it is waste. Both of these ideas have worked well in Europe but not in Aboriginal territories, where people did not (a) unilaterally "work" the land without acknowledging it or (b) consider any part of land as "waste." Land, it is to be remembered, is considered a living entity known to the ancient Nuu-chah-nulth as Ḥaw'iⱡume, or Wealthy Mother Earth, and, consequently, no part of this land could possibly be *waste*. Appropriately, then, a Nuu-chah-nulth national territory was subdivided into *haḥuuⱡi* (land owned by a chief). Anything of value found within a specific *haḥuuⱡi* was brought for a finder's fee to the chief of that particular *haḥuuⱡi*. The first white settlers on the coast, according to common oral history (which includes stories that I heard as a child), understood this convention and obeyed it by bringing things of value to their proper owners.

The problems of land claims in Canada were created, in part, by the peculiar ideas, propagated by philosophers like Locke, about land and how title to it is acquired. These ideas, which included the erroneous conception that Aboriginals were childlike, instinctive, savage, and barbaric, were translated into the provisions of the Indian Act. It is for this reason that it was thought to be unnecessary to get Aboriginal consent regarding where First Nations peoples would live, how they would live, how and where they would be educated, and whether they could be considered to be fully human. These erroneous perspectives on Aboriginal peoples add to the difficulty of interpreting Aboriginal rights today. Politicians, lawyers, judges, philosophers, academics, researchers, and leaders have all been trained within an educational system that has, *until very recently,* been completely biased in its representations of Aboriginal histories and cultures. This educational system, as indicated by a study published in 1958 by H.B. Hawthorn, C.S. Belshaw, and S.M. Jamieson, led to a general belief that Aboriginals were of a low intellectual capacity and without "the potential to develop as rapidly as Whites along the lines of social, emotional, educational, moral or economic attainment."[24]

Therefore, when the potlatch was outlawed in 1884, no one in government thought to ask any of the Aboriginals whether they had an opinion on this attack on their traditional form of government. No one thought to ask whether Aboriginals had any institutionalized cultural correctives that might take care of socio-political problems.

6

The Nuu-chah-nulth Principle of Continuity

Iisʔak̓, sacred respect, is consistent with the Nuu-chah-nulth assumption that all life forms have a purpose.[1] The existence of life forms is given value by K̓ʷaaʔuuc, Owner of All. Value exists by association with K̓ʷaaʔuuc and, thus, all life forms have value and all are to be allowed to continue to live sustainably because of this value. Consequently, continuity is a fundamental tenet of Nuu-chah-nulth constitutionalism, which was set into practice through the development of protocols between peoples, nations, animals, and other life forms.

Although *iisʔak̓* is of ancient origin, it has never been articulated across cultures in a way that has been cognitively understood, particularly in legal or constitutional terms. Typically, among traditional indigenous societies, such as the coastal Nuu-chah-nulth and Salish peoples, the interpretation of *iisʔak̓* is not an issue. This point is critical because *iisʔak̓* is perfectly consistent with Amerigo Vespucci's sixteenth-century observation that the New World Aboriginal is "lord of himself." Since each is lord of her- or himself, each, within the bounds of accepted cultural beliefs, can make her or his own decisions. This lordship of self involves two streams of perception: that of the self (the micro) and that of the group (the macro). It is for this reason that an origin story, such as the one about Son of Raven, when taken from either an individual or a community perspective, is, in principle, the same story even though it may be interpreted differently by each person. In general, what this means is that *iisʔak̓* is recognized as a potential principle of life, which is defined as broadly as the imagination of the total number of community members allows. Consequently, just as each is lord of him- or herself, so too does each have his or her own way of practising *iisʔak̓,* his or her own understanding, all of which, taken together, make up the accepted beliefs and customs of a people's lifeways. Unlike other ways of interpreting the spiritual domain, the ancient Nuu-chah-nulth made no attempt to restrict this domain to a specific definition or to confine its understanding to an elite school of philosophical thought.

What indigenous cultures mean by the "value of life forms" can only be assumed, not known, by Western cultures. In other words, as soon as the phrase "value of life forms" is applied across cultural divides, a problem of interpretation arises. For example, Jean-Jacques Rousseau's notion of Aboriginals implies a very different interpretation of this phrase than that held by a traditionally oriented Nuu-chah-nulth person. For the ancient Nuu-chah-nulth, it involved the belief that plants, animals, and people were all alike in that each could communicate with the other in an intelligible fashion, while Rousseau imagined Aboriginals to be purely instinctive, without laws and morals and consequently unable to communicate effectively with each other. The crux of the issue is not what Rousseau meant or intended by his writings but,[2] rather, the way in which his writings were broadly interpreted and then employed to justify and support the colonial project.

A Nuu-chah-nulth Meaning of Continuity[3]

Given the historical circumstance of the meeting between those of European extraction and those indigenous to the Americas, and given that indigenous ways of life were authoritatively and prevailingly misunderstood by the colonizers, it may no longer be a surprise that their coexistence continues to be characterized by a wall of mutual misunderstanding. What, then, does continuity mean to traditional Nuu-chah-nulth?

Continuity at the Personal Level

In the story of Son of Raven and Eagle we find a violation of personal boundaries and personal integrity. Son of Raven observes a powerful, magnificent, and spectacular process of fishing as Eagle dives swiftly but silently to capture a salmon from the river. In his comic enthusiasm, Son of Raven thinks to do the same and foolishly invites guests to a great salmon feast. He then flies high into the sky in the same way that he observed Eagle do. There, far below in the river, he sees a dark spot. Ah, ha! It must be an unsuspecting salmon! He begins his descent and gains momentum on the way. Crash! He knocks himself unconscious and floats unceremoniously down the river with his feet in the air. All Nuu-chah-nulth know that Son of Raven, unlike Eagle, has poor eyesight, and the dark spot in the river turned out to be a reef.

Eagle has a way of life, as does Son of Raven. When the latter attempts to imitate the former, he discontinues his way of life. Another way to put it is, Son of Raven momentarily forgot his own identity. From this illustration, the meaning of continuity does not imply that everyone must be the same or that everyone must be identical. Each must continue as a diverse being. The continuity of a way of life at the personal level is a major constitutional theme in every traditional Nuu-chah-nulth household.

Continuity across Apparent Divides

In *Tsawalk: A Nuu-chah-nulth Worldview,* I discuss how the story of Son of Raven and his community can also be about the violation of personal boundaries and personal integrity. By transforming himself into a larger salmon than do his community members, he distorts his own personal image relative to the personal image of his fellow community of beings. Son of Raven fancies himself greater than others of his kind. Moreover, this distortion of self seems to have consequences that are so consistent as to suggest a natural law. For the purpose of this argument, this natural law may be defined by the need to have *integrity of being.* The journey to attain integrity of being is a developmental process, a maturational process, and a process of discovering the integrative nature of creation.

Son of Raven, it turns out, is a created being among other like-created beings who, in a purposefully challenging and polarized creation, have equitable value and are potentially able to live together in a mutually sustainable way without violating personal boundaries. The process towards the possibility of sustainable living for all life forms is necessarily developmental, which naturally requires the application of the characteristics that define a healthy person – strong, visionary, caring, loving, cooperative, creative, insightful, discerning, empathetic, hopeful, faithful, and enduring. It is this sense of unity with the whole of reality (a natural purpose of existence) that seems to speak to an *ultimate constitutional right to continuity.*

Son of Raven has the natural right to continue his way of life as Son of Raven in a community and as an heir of the Creator rather than as a poor copy of someone else.

When we know ourselves, that we each have our own integrity of being, that Raven is Raven and not Eagle; when we perceive our relative size and range of power compared to creation and Creator; when we recognize the possibilities of creation within this limitation; when we are realistically oriented to the nature of reality, even if imperfectly understood, even if our knowing is more a sense of things than a clear understanding; then we begin to grasp some of the mystery of creation and its purpose.

During the nineteenth century, my great-grandfather Keesta had his own unique ?uusumč practices, in specific places, at specific times of the year, and with specific items of ritual. As a direct result of the residential school intervention I learned none of Keesta's specific practices, but I did learn his underlying belief system and do know many of the common items used for ritual, such as the cedar and hemlock boughs. As a result, without specific training or guidance, I developed my own ways, and they have proven effective. For example, I have developed a contemporary ritual using four branches of a certain tree. My mother called me on the phone one day and asked for help. For some weeks, in spite of her best intentions and in spite of a need to make a little money on the side, she had been unable to work on her traditional grass baskets. Since she was otherwise in good health there seemed no explanation for her inability to work other than the existence of negative spiritual activity. I performed my ritual with my four branches and the next day my mother called me to say, with some excitement, that she now had no problem working with her baskets. This story is meant to illustrate one small example of the intergenerational continuity of a cultural practice that did not depend solely upon human-to-human transfer of knowledge but, rather, entailed a dynamic transfer between the physical and non-physical domains of existence. The ritual items I employed were common to Nuu-chah-nulth, but the specific way in which I performed the ritual differed from the way of my grandparents.

For all those who have a different worldview from that of the Nuu-chah-nulth – particularly those who have grown up under Darwin's theory of evolution, which assumes reality to be governed by random mutation and the survival of the fittest – my attempt to describe this traditional way of life may continue to be problematic. For example, if everyone is allowed to be different, to do as he or she pleases in terms of important cultural institutions such as the *ʔuusumč,* is this not a description of anarchy, of disorder, of chaos, of an absence of good government? In addition, how is one to know the truth of things if each time the story of Son of Raven and his community is told some of the details change? Worse still, the details may vary widely from community to community and from nation to nation.

One answer to these questions can be illustrated by an analysis of music. A piece of music can have many variations and interpretations. Yet, in spite of variations in the way specific notes are played, the musical theme remains the same. Another answer is found in dance. Each dance of life may be done to the same music, but each can also be unique. Yet, in spite of multiple interpretations of each piece of music or dance, the original musical score or dance routine can be identified. Moreover, from the perspective of human development, a unique and personal dance of life implies integrity of being, which is important to personal health and identity. The metaphors of music and dance can illustrate continuity not only within Nuu-chah-nulth culture but also across the apparent divides created by different worldviews. That each person has his or her own way of life, his or her own view of reality, his or her own worldview, does appear to create divisions between people. However, from a Nuu-chah-nulth perspective, if these unique lifeways are examined more closely, each person can be found to be dancing to the same kind of music. Each person believes that a *ʔuusumč* is important to survival, to being informed, to having food on the table and medicine in the cabinet. Yes, each story may have different details (just as each song may have different notes), and yes, each *ʔuusumč* may vary from person to person, but the underlying theme of life, to which all ancient Nuu-chah-nulth danced, has proven both indispensible and indestructible – so much so that the legislation, policy, and practice of colonial governments could not entirely extirpate it.

Continuity in, and of, Nature

It may be recognized that the need to articulate a constitutional principle of continuity – the sacred right of life forms to continue to live in their own integrity of being – is a co-management principle. This co-management principle is the same as is Elisabet Sahtouris's idea of mutual consistency.[4] If the great diversity of life forms in nature is interconnected and dynamic and it is to continue to survive, then these life forms must learn to co-manage their common reality. Survival, in this sense, is, metaphorically speaking, like harmonious music. For Sahtouris, "mutual" means shared and "consistency" means agreement or harmony.[5] A mutually consistent biodiversity ensures continuity.

In the context of a reality characterized by polarization, by dichotomies and contradictions that often appear chaotic, we find in Nuu-chah-nulth stories that Wolf has hammered out an agreement, a protocol, with Deer. We find that Salmon has hammered out an agreement, a protocol, with the Nuu-chah-nulth. Keesta hammered out an agreement, a protocol, with three whales. Son of Mucus had come to Ahous to rescue children that had been abducted by Pitch Woman and then returned to his spiritual home to marry the chief's daughter. Son of Mucus took for granted that there were protocols to observe before any marriage (an example of mutual consistency) could take place.

All protocols take place within a context that appears contrary to the spirit of agreement and harmony, contrary to the constitutional principle of continuity. Predation – by Wolf upon Deer, by colonizers upon colonized, by corporations upon the powerless and poor – is a common reality and is therefore a dimension of truth. The Nuu-chah-nulth discovered that predation is natural to creation, with or without management. Without purposeful agreements between the diverse inhabitants of reality, danger, destruction, conflict, and hatred remain inevitable. In other words, according to ancient Nuu-chah-nulth, life is dangerous if it is not appropriately managed but beautiful if it is appropriately managed. It is for this reason that ancient Nuu-chah-nulth observed protocols in their "predation" upon any plant or animal. This type of predation, which is conducted through consultation and agreement, results in taking what you need rather than in taking what you don't need. Need, when appropriately fulfilled, results in health, harmony, sustainability, and continuity. In this sense, taking what you need is a law of sustainability and an example of the constitutional principle of continuity.

Then Came Democracy

John Dunn makes this pithy comment about democracy: "As an expectation about the human future it is little better than absurd."[6] Although the word "democracy" has a long history, it did not always enjoy the stellar reputation that it has today. "What is very strange indeed (in fact, quite bizarre)," Dunn writes, "is the fact that this single term, endlessly transliterated or translated across all modern languages, should turn out to be the ancient Greek noun *demokratia,* which originally meant not a basis for legitimacy, or a regime defined by its good intentions or its noble mission, but simply one particular form of government, and that a form, for almost two thousand years of its history as a word, which, it was overwhelmingly judged by most who used the term, had proved grossly illegitimate in theory and every bit as disastrous in practice."[7] Dunn indentifies the word's later usage: "With the French Revolution, democracy as a word and an idea acquired a political momentum that it has never since wholly lost. Only after 1789, as far as we know, did any human beings begin to speak of *democratizing* the societies to which they belonged."[8] It was slightly more than two hundred years ago, then, that the idea of democratizing *societies* began to take hold. Given democracy's youthful and emergent state of being, and given its stellar reputation, we are reminded by Francis Fukuyama that: "The apparent number of choices that countries face in determining how they will organize themselves politically and economically has been *diminishing* over time. Of the different types of regimes that have emerged in the course of human history, from monarchies and aristocracies, to religious theocracies, to the fascist and communist dictatorships of this century, the only form of government that has survived intact to the end of the twentieth century has been liberal democracy."[9] Can Fukuyama's statement be taken at face value or are the values of old monarchist and colonial regimes simply finding new ways to express themselves under the liberal democratic tradition?[10]

If liberal democracy has had a problem with the constitutional principles of recognition and consent as these apply to women, it has certainly had a problem with the principle of continuity as it applies to indigenous populations as well as to plants, animals, and the living environment. The principle of continuity as it applies to the right to human life was first expressed in

1997 by Chief Justice H.E. Hutcheon in the Supreme Court of Canada in the *Delgamuukw* case. Hutcheon's statement is a complete reversal of the early colonial notion that "the only good Indian is a dead Indian." The Supreme Court record reads as follows:

> I develop this point with respect to the test for aboriginal title. The relevance of the continuity of the relationship of an aboriginal community with its land here is that it applies not only to the past, but to the future as well. That relationship should not be prevented from continuing into the future.
>
> ...
>
> Ultimately, it is through negotiated settlements, with good faith and give and take on all sides, reinforced by the judgments of this Court, that we will achieve what I stated in *Van der Peet,* supra, at para. 31, to be a basic purpose of s. 35(1) – "the reconciliation of the pre-existence of aboriginal societies with the sovereignty of the Crown." Let us face it, we are all here to stay.[11]

"Let us face it, we are all here to stay" is one way to express the constitutional principle of continuity. Although this phrase refers exclusively to human relationships, it can, from an indigenous perspective, be generalized to include all life forms – plants, animals, and humans. An indigenous interpretation of the phrase would take it to mean sustainability, or the maintenance of balance and harmony, between various life forms. This is not the case from the Government of Canada's perspective, however, as, for one thing, the Indian Act presents a major obstacle to such an interpretation. Nevertheless, the Supreme Court declares, with hope and some understatement, that the difficult task is to "achieve ... the reconciliation of the pre-existence of aboriginal societies with the sovereignty of the Crown," which is the basic purpose of Section 35(1). Without a balanced and well thought out historical understanding between Canada and its Aboriginal population, the interpretation of Section 35(1) is likely to remain an issue for some time.

In Canada there are two interrelated problems with the principle of continuity, and these remain issues for its Aboriginal population: (1) the problem of colonial history and (2) the problem of an emergent, and perhaps immature, state of liberal democracy.

The Problem of Colonial History

In his book *Native Roots: How the Indians Enriched America,* Jack Weatherford writes:

> Even though today we no longer share Columbus's folly of thinking that we are in Asia, we still do not adequately know where we are. We have built cities and cleared farms across the continent, but we do not know the story of the land on which we live. We take nourishment from this soil, but because we cannot see our roots down deep in the American dirt, we do not know the source of that nourishment.
>
> Our cultural roots as a modern people lie buried in Cahokia and a thousand similar historical sites and surviving Indian reservations across the continent. These ancient and often ignored roots still nourish our modern society, political life, economy, art, agriculture, language, and distinctly American modes of thought.[12]

Since the "story of the land" has been told primarily from the colonizer's perspective, the *full* story remains hidden. The official version of the history of North America began at contact – specifically, when Columbus landed in 1492. In the schools of North America this history is well rehearsed, but the story of the land before this time remains generally unknown because, until very recently, it has been omitted from the school curriculum. This created an educational vacuum in mainstream curricula, as affirmed by Thomas Berger:

> In 1963 when I opened my own law practice in Vancouver, I had few clients. Among the few were two Indians from Nanaimo, Clifford White and David Bob, charged with hunting deer in the closed season ... When I had gone to law school in the 1950s, the idea of aboriginal rights had never been discussed. Law schools paid no attention to the issues of

Indian land or Indian rights. It was not even a marginal subject ... No one studied it, and no one thought about it ... We won the case on the ground that the Indians at Nanaimo had entered into a treaty in 1854 recognizing their hunting rights during the closed season.[13]

The significance of Berger's educational experience is that it is generalizable to all Canadians during most of the twentieth century. At the ninety-fifth birthday party of Walter Koerner,[14] the former British Columbia Supreme Court judge Allan McEachern told me that, given the schooling he had received, his decision against the Gitksan-Wet'suwet'en in the *Delgamuukw* case could not have been avoided. Like Thomas Berger, McEachern did not study the issues of Indian land and rights because such studies were unavailable; instead, he studied various Enlightenment authors and based his decision on their ideas about Aboriginals. These authors, such as Thomas Hobbes, Jean-Jacques Rousseau, and John Locke, continue to be studied for their contributions to political science, including their misguided views about Aboriginality. A related example can be found in an account provided by the Honourable A.C. Hamilton, who said:

Although I attended what I assume were good schools – Grosvenor, Robert H. Smith and Kelvin – as far as I can recall there was little in my education that dealt with Aboriginal history or the part Aboriginal people had played in the development of Canada. I certainly never received any education about Aboriginal people. We learned the rhyme: "In 1492 Columbus sailed the ocean blue" when learning important dates in history, but that, and the names of his three ships, was the extent of my understanding of that event or its impact on the Americas.[15]

The educational experiences of Hamilton and Berger may be considered typical. Canadians – including prime ministers, premiers, mayors, Supreme Court judges, presidents of universities and colleges, school principals, most professors and teachers, business leaders, and those in the media – can say: "I certainly never received any education about Aboriginal people."

The Emergent State of Liberal Democracy

According to John Dunn, the idea of democracy has had legitimacy only since the time of the French Revolution. That is a short period in which to develop a complex way of life. As a consequence, since its beginning, the application of democracy has been severely limited. Of the four constitutional principles of ancient Nuu-chah-nulth society – recognition, consent, respect, and continuity – only consent prevails as both a constitutional principle and a general democratic practice.

The democratic principle of continuity, which addresses the fundamental issue of the right of life forms to continue their ways of life, does not formally exist within the constitutions of the Western world. Rather, there is a doctrine of continuity that functions as a legal principle for identifying precedent, and there is also a notion of continuity whose meaning is restricted to an orderly transfer of governmental powers from one party to another. In the latter case, ensuring continuity is merely a political act of self-preservation – an act that is reflected in the second half of the title of Darwin's famous work (*Preservation of Favoured Races in the Struggle for Life*) and that leaves the vast majority of life forms without legal remedy. During much of the short history of liberal democracy, the practice of another democratic idea closely related to continuity – namely, equality – invites the same kind of interpretation.

At the rational level, one would think that the cry of "liberté, égalité, fraternité" shouted by those involved in the French Revolution might have ushered in the beginning of a struggle towards balance and harmony – at least between humans of similar skin and eye colours if not between humans from different cultures and worldviews. The fact that the French Revolution was followed not by increasing measures of liberty, equality, and fraternity but, rather, by more conflict, inequality, and division only highlights the difficulty of realizing democratic ideals. As a child in grade school I was completely misled by the curriculum, which taught that Canada was a free country. On the playground, whenever another boy attempted to direct my conduct, I would always say, "You can't tell me what to do, this is a free country." The irony of this statement is that it was made within the grounds of a residential school, where we were physically punished if we strayed outside its fenced boundaries. It has taken half a century for me to realize that much of the material that I learned in school did not apply to me. I lived my life according to a sort of Orwellian double-think: I believed that I was part of a free country of equals, yet at the same time I was keenly aware that, from the 1940s through the 1970s, I was refused service in most restaurants and most motels and hotels simply because of my aboriginality.

This kind of experience of inequality, shared by many people in any free country, suggests that liberal democracy remains in a stage of life that, for purposes of illustration, we might associate with kindergarten. The primary goal of kindergarten is to teach little children how to get along with one another (the constitutional principles of recognition, consent, continuity, and respect). In this first stage of formal education, children are taught to share their toys (political goods) and not to fight over them (wars). The educational goals for kindergarten students parallel the goals of a liberal democracy. The one great difference between the two is that, while the former have an authoritative teacher to guide them, the latter does not.

One of the many examples of this early stage of constitutionalism in Western democracy is the St. Louis World's Fair, held in 1904, at which fifty-nine Aboriginal groups from around the world were put on public display as representatives of early life forms on Darwin's evolutionary scale.[16] Five Nuu-chah-nulth were among the display, Atlieu, an ʔuuštaqyu (shaman/doctor); his daughter Annie Atlieu; his half-brothers Jasper Turner and Jack Curley; and his mother Ellen Curley. To convince Atlieu that he should go to St. Louis, one of the fair's organizers, indicated to him that one purpose of his visit would be to meet other great leaders like himself. The Canadian historian Douglas Cole explains the rationale for the St. Louis World's Fair as follows:

WJ [sic] McGee's aim in the fair's anthropology section was "to satisfy the intelligent observer that there *is* a course of progress running from lower to higher humanity, and that all the physical and cultural types of man mark stages in that course." To assist in demonstrating this evolutionary model he brought to St. Louis African pygmies, Ainu from Japan, Tehuelche from Patagonia, and, among American Indians, representatives from the Cocopa, Pawnee, Dakota, Pueblo, Pima, and Pomo groups.[17]

Consistent with this Darwinian logic, the colonial project exercised the full force of state policy, legislation, and practice in a concerted effort to *discontinue* First Nations lifeways in Canada. The consequences have been comprehensively documented both by the Hawthorn Report of 1966-67 and by the Royal Commission on Aboriginal Peoples of 1995-96. For example, psychologist Michael Chandler and colleagues, in a study of suicide, identify First Nations youth in Canada as a group who "take their own lives at rates that are said to be higher than that of any culturally identifiable group in the world."[18] They continue:

But where, you might well ask, is the surprise, let alone a puzzle, in that? Who now belatedly fails to understand that it is really we who savaged them? In the wake of centuries worth of genocidal practises and publicly endorsed programs of ethnic cleansing, few are likely to be especially caught off guard on learning that many Canadian and American aboriginal youth simply judge life no longer worth living. No, the real surprise and the real questions arise in response to the mind-numbing size of the actual body count – the sheer amount of blood on the floor. How, we want to know, did things get so far out of hand? How did it come to pass that, for so many, death is the preferred alternative? How, despite what are meant to pass as contemporary good intentions, did we collectively manage to shoulder our way past places such as Bangladesh or Rwanda, or even Afghanistan (places that, as we imagine them, seem even more beyond hope of decent human prospects), to capture the number-one spot in this dark competition?[19]

Although the foregoing is now well-documented history, such information remains marginal. John Ralston Saul puts this another way: "When non-Aboriginals are being honest with themselves, they admit that the denial of indigenous culture destabilized the Aboriginals. This is accurate."[20] Here, Saul is saying that, when Canadians deny this significant aspect of Canadian history, they also deny the fact that: "What we are today has been inspired as much by four centuries of life with the indigenous civilizations as by four centuries of immigration ... Today we are the outcome of that experience."[21] And Aboriginal suicide rates are part of that experience. Aboriginal suicide rates are a significant part of what Canada is and continues to be. Although Canada's Aboriginal history may be denied by some, it is also true, as I've indicated in the cases of Thomas Berger and A.C. Hamilton, that this denial is based on an absence of information about Aboriginals in the school curricula.

Either because of a lack of good information or because of a deliberate denial of Canadian history, which destabilized the Aboriginals, the Supreme Court decision in *Van der Peet* demands that any claim to Aboriginal title be tied to Aboriginal identity:

> In *Van der Peet,* I drew a distinction between those practices, customs and traditions of aboriginal peoples which were "an aspect of, or took place in" the society of the aboriginal group asserting the claim and those which were "a central and significant part of the society's culture" (at para. 55). The latter stood apart because they "made the culture of that society distinctive ... it was one of the things which truly made the society what it was" (at para. 55). The same requirement operates in the determination of the proof of aboriginal title.[22]

The Supreme Court of Canada holds the position that to maintain an identity described as "distinctive," Aboriginal societies must be unchangeable. In effect, the judicial system, which, through the Indian Act, directly contributed to the erasure of Aboriginal identity and the forceful distortion of First Nations lifeways, now demands that any claim to land title be based, in part, on these same lifeways. On this particular point, Neil Vallance, lawyer and anthropologist, writes:

> My survey found no cases, outside the area of Aboriginal rights, in which claimants were required to prove anything about their culture as a prerequisite for entitlement to rights. In other words, claimants to Aboriginal rights are held to a higher standard than claimants to Charter or other

rights. I believe that the creation of this double standard constitutes an injustice against the First Nations in Canada.[23]

Is Continuity Possible?

This is a reasonable question, given the long-standing misunderstanding on many fronts, including the legal, educational, political, economic, and spiritual. In spite of its apparent impossibility, the answer is definitely, "Yes, continuity is possible." In my own community of Ahousaht, during my childhood our living oral history was still being passed on by grandparents and great-grandparents who, towards the end of the nineteenth century and the beginning of the twentieth, had had first-hand experiences with new settlers to the west coast of Vancouver Island.

Ancient laws related to chiefly rights known as *hahuuɬi* were routinely observed during this time. As indicated earlier, one of the laws of *hahuuɬi* concerned items of value found within the territory of a chief. If a canoe, for example, had drifted away during a storm and landed on some beach owned by a chief, the law of *hahuuɬi* stated that it must be brought to the owner of the land, who would pay a fee to the finder. The first white settlers in the area understood this practice and complied with it. This is an example of mutual understanding, whereby newcomers recognized the need to respect indigenous laws. The historian Robin Fisher explains the basis of this mutual understanding:

> During the fur-trading period Indians and Europeans shared a mutually beneficial economic system. Because the fur traders were involved in an enterprise that required the co-operation of the Indians, they did not try to alter their social system radically, to undermine their beliefs, or to destroy their means of livelihood. Indeed, fur traders, at least to some extent, had to accommodate to Indian ways.[24]

The fur traders did not come to take away the land or to do away with the languages and cultures of the indigenous peoples. They came to buy and sell and, in so doing, discovered that their indigenous trading partners were as skilful as any in the Old World. There was occasional violence and misunderstanding on both sides, but this is not unusual between traders and between those who do business with each other.

Towards Mutual Understanding

The accounts of the initial encounters between Indians and Europeans are windows into the past, even if the glass is smeared and distorted by the chroniclers' prejudices and misapprehensions.

– Charles C. Mann, award-winning correspondent for the magazines Science *and the* Atlantic Monthly[25]

Charles C. Mann's *1491: New Revelations of the Americas before Columbus* clears away some of the erroneous perceptions – created by explorers like Columbus and Vespucci and perpetuated by the writers of the Enlightenment – that, at contact, the Americas were a vast wilderness. They were not a vast wilderness but, rather, well managed and massively landscaped. According to Mann: "Indians retooled ecosystems to encourage elk, deer, and bear. Constant burning of undergrowth increased the numbers of herbivores, the predators that fed on them, and the people who ate them both." Moreover: "Indian fire had its greatest impact in the middle of the continent, which Native Americans transformed into a prodigious game farm. Native Americans burned the Great Plains and Midwest prairies so much and so often that they increased their extent; in all probability, a substantial portion of the giant grassland celebrated by cowboys was established and maintained by the people who arrived there first."[26]

Although burning vast tracts of land appears to be destructive, it is actually part of a healthy process of ecosystem management. The Nuu-chah-nulth also practised controlled burning to promote a more productive ecosystem. In my view, this treatment of nature is also a principle of *ʔuusumč,* in preparation for which a man may rub branches on his body until he bleeds. He may dive into a pool or lake and stay under water so long that his nose bleeds. He may subject his body to such extreme privation that he may bring himself to the edge of death. This behaviour may be described as pathological. However, it is necessary for him to reach these extremes as previous experiences indicate that this is where he may find the strength, power, and spiritual gifts that will enable him to procure the good things of life. Controlled burning brings an ecosystem to the edge of death so that it can produce the good things of life – an abundance of berries, an increased herbivore population, and an abundance of food for humans.

This similarity between the process of controlled burning and *ʔuusumč* can be illustrated through an analysis of *tsawalk,* one. The unity of creation is filled with living beings who are characterized by an integrity of being that defines the boundary between each of them. In the *ʔuusumč* the boundary is between the physical domain and the non-physical domain, which, in the beginning, was violated by Son of Raven. After a process of trial and error, the ancient Nuu-chah-nulth learned how to respect the integrity of the boundary between the physical and the non-physical. This boundary is found primarily within the person, the *quuʔas*. Becoming, metaphorically, like an insignificant leaf involves purposefully shifting internal personal boundaries between the ego of self and the spirit of self. This is a description of polarity, of inner conflict, that is found naturally within any given being. In the management of this internal polarity, the *ʔuusumč* practitioner assumes a humble stance. A state of humility enables the placement of the boundary between the physical and non-physical domains so as to maximize the intended dynamic relationship between Creator and created. This is not a description of humiliation or debasement but, rather, an acknowledgment of the relative insignificance of the human in comparison to a creation that remains mostly a mystery. Any distortion of the boundary of self that places it beyond the boundaries of this insignificance must, by definition, violate the principle of continuity. The same can be said for the practice of controlled burning. The factor of "control" implies that the purpose of the act of burning is to promote the growth of food to benefit both humans and animals. An out-of-control forest fire, having no boundaries, is destructive to both humans and animals.

What is said about the *ʔuusumč* and about the controlled burning of an ecosystem and their relationship to the principle of continuity can also be applied to the boundaries between nations and cultures. Following this, what Mann does for indigenous peoples from other parts of the Americas, I am attempting to do for the Nuu-chah-nulth peoples of the Northwest

Coast. In 1792, from 29 April to 21 September, José Mariano Moziño, a botanist-naturalist given to scientific observation, made journal notes about the Nuu-chah-nulth:

> The government of these people can strictly be called patriarchal, because the chief of the nation carries out the duties of father of the families, of king, and high priest at the same time. These three offices are so closely intertwined that they mutually sustain each other, and all together support the sovereign authority of the taises. The vassals receive their sustenance from the hands of the monarch, or from the governor who represents him in the distant villages under his rule. The vassals believe that they owe this sustenance to the intercession of the sovereign with God. Thus the fusion of political rights with religious rights forms the basis of a system which at first glance appears more despotic than that of the caliphs and is so in certain respects, but which shows moderation in others.[27]

To add credibility to his observations, Moziño wrote that he was able "to learn about the various customs of the natives, their religion, and their system of government" because he *"learned their language sufficiently to converse with them."*[28] To demonstrate his knowledge of the language, Moziño provided a "Brief Dictionary of the Terms That Could Be Learned of the Language of the Natives of Nootka." Since Nuu-chah-nulth (formerly Nootka) is my first language, I learned as a young child that K̓ʷaaʔuuc is equivalent to what the English-speaking Canadians refer to as "God." Much later, at university, I came across this work by Moziño in which the word "God" is translated phonetically in the Nuu-chah-nulth language as "Coautz," or, in the current orthography, K̓ʷaaʔuuc. It appears, since we still have a number of fluent speakers, that the Nuu-chah-nulth language has survived, or, more accurately, continues to exist, although in limited form.

Moziño's description of precontact Nuu-chah-nulth lifeways is the more remarkable because he, as an outsider with a scientific worldview, was able to classify in broad structural terms the relationship of these Aboriginals to their land. Their government, described as patriarchal, is divided into three major areas: the social (father of the families), political (king), and spiritual (high priest). In addition to the explicit description of these three major areas of governance, there is also a reference to the responsibility of those who are referred to as *taises* (singular *taise,* or *tyee*), one literal translation of which is "eldest." It is the responsibility of the *tyee* to provide for the "sustenance" (the economic area of governance) of the general populace.

The Office of High Priest

Moziño was somewhat correct in his assessment of those whom we today call the head chiefs. However, no chief was ever like a high priest in the sense that he would have guided other priests for, as Vespucci correctly observed, each person was "lord of himself."[29] That is to say, each person, who was a member of a nuclear family in an extended household, was a priest in the same sense that the *ʔuusumč* was a universal practice. There were no universal codes of conduct for the priestly role as each practitioner of the *ʔuusumč* either developed his or her own code or followed a code handed down from a direct ancestor. Even when a code of conduct for a *ʔuusumč* was handed down, it was likely to be modified. Moziño was correct in describing the head chief as a high priest in so far as the former was, from time to time, expected to provide for the entire community, either directly through his own practice of the *ʔuusumč* or indirectly through the resources of his *haḥuułi*.

The Office of King

Here again, Moziño was only somewhat correct, for the head chief was indeed like a king, but he did not have the absolute powers that often accompanied the rule of European monarchs. In fact, Moziño wrote in his journal that ordinary citizens often related to Maquinna, who was the head chief of the Muwachat, in a casual manner rather than with the obsequiousness demanded by kings and queens from other continents. The relationship between *musčim* (ordinary citizens) and the head chief could be measured by the degree of success with which the *ʔuusumč* was practised. A successful *ʔuusumč* always translated into the provision of sustenance for the community. Another avenue of success was the natural resource wealth of a *haḥuułi*. If a head chief owned resource-rich territory, and these resources were parcelled out judiciously so that the entire community was satisfied, then this was always acknowledged in the great ceremonies of the feast halls. Moziño does not write about formal council meetings in his journals, no doubt because he was never invited to one. Each chief had councillors, or advisors, who were responsible for discussing and executing all matters relating to what might be considered the state.

Nuu-chah-nulth Relationship to Land

Moziño makes no mention of the special relationship that the Nuu-chah-nulth had with their land. Just as, in 1997, the Supreme Court of Canada could not define the specifics of Aboriginal peoples' relationship to land, so in 1792 Mozino was unable to define the specifics of the Nuu-chah-nulth relationship to land. What he does describe is an economic relationship between a Nuu-chah-nulth leader who is said to be like a *father, king,* or *high priest,* who provides *sustenance* for the general populace. Much of what today is called a special relationship to land has now been usurped by environmentalists, some of whom strive to emulate and even appropriate what they believe to be Aboriginal attitudes towards nature. For example, environmentalists on the west coast of Vancouver Island may describe the temperate rain forests as being like a cathedral or a temple. These descriptions are somewhat correct, in the same limited way as describing a head chief as a high priest or king.

But even the phrase "special relationship to land" is a distortion of Aboriginal views, a point well illustrated by a question that was asked by scientists on the Scientific Panel for Sustainable Forest Practices in Clayoquot Sound: "Where are your sacred sites?" This question assumes that "sacred" is separate from what those in the Western world call "profane."

The Nuu-chah-nulth relationship to land is implied in the term "Ḥaw'i-łume," which means "Wealthy Mother Earth." A mother is recognized for her ability to bear children and to nourish them from her breasts. Wealthy in all ways, physically and spiritually, she provides an abundance of physical nourishment while she cooperates spiritually in the provision of her branches, leaves, and roots so that, in the sacred practice of the *ʔuusumč,* the intended unity of all things physical and spiritual can be realized. From the point of view of governance, national boundaries, and the egalitarian distribution of political goods, the Nuu-chah-nulth relationship to land is bound up in the sovereign notion of *haḥuułi*. In this one word can be found a metaphor for creation's intended purpose and unity – *ha* (that which is empirically close and observable), *ḥuu* (that which is hidden and beyond), and *ł* (the land upon which *ha* and *ḥuu* are ceremonially brought together through the practice of the *ʔuusumč*).

The Educational Factor

Between 1632 and 1972, First Nations students performed extremely poorly, as was guaranteed by policy and practice. However, the very fact of this massive failure creates some hope for Aboriginal lifeways. As I have pointed out, early European methods of education proved so alien to First Nations students that, in 1644, Mother de l'Incarnation reported: "We find ... intelligence in them, but when we are least expecting it they climb over our enclosure and go to run the woods with their relatives, where they find the more pleasure than in all the amenities of our French houses."[30] It is ironic that the combination of (1) educational methods designed for cultural genocide and (2) the constitutional ban on First Nations cultural practices resulted in Aboriginal peoples' stiffening their resistance and taking their cultural practices underground.

Now that educational methods have changed to become more inclusive of First Nations lifeways, and given that the constitutional ban on the potlatch was rescinded in 1951, it has become possible to revisit and to clarify historical misunderstandings created by colonial hegemony. At present, however, at the postsecondary level Aboriginal curriculum remains marginal. The Canadian K-12 (kindergarten-to-Grade 12) curriculum has two main streams: core and supplementary. As of 2009, virtually no Aboriginal studies courses fall within the core program that all pupils must complete. This exclusion of Aboriginal content from the compulsory curriculum is beginning to change, and this will help to begin to remedy the omissions of the past. And this may, in turn, make possible the realization of the principles discussed in these chapters.

7

Haḥuułism

"Our Stories Are True!"[1]

My goal and vision for this chapter is to begin to explore what is meant by *haḥuułism*. Although *haḥuułism* is a neologism, its meaning is as ancient as creation itself. Most "isms" tend to describe a philosophy of life. Socialism, communism, liberalism, capitalism, and even environmentalism are all descriptions of a way of life. In so far as each "ism" describes a way of life, whether communal or individual, socialistic or capitalistic, it is of the same descriptive class as *haḥuułism,* a word that I have created and use to describe a traditional Nuu-chah-nulth way of life, of which only vestiges remain in story, song, chant, dance, regalia, language, and practice. Again, it must be made clear that, even if "our stories are true," this does not mean that they continue to be translated meaningfully into everyday lives. The project that I identify as *haḥuułism* was interrupted and put on hold during the colonial era but can now re-emerge and be translated so as to be of some use during this time of global crisis.

But before I discuss this new word, I will attempt to show why its introduction may add not just to Nuu-chah-nulth knowledge but also to human knowledge in general. *Haḥuułism* may add to Nuu-chah-nulth knowledge because its principles are taken from ancient beliefs and practices that sought constantly to strengthen life through emphasizing relationships between all life forms. In itself this emphasis on strengthening life represented a gargantuan struggle that was difficult to win and, when won, was equally difficult to maintain. While knowledge of the cultural process of this daily struggle survives in the modified form of the current practice of the *ʔuusumč,* it is now evident that, when juxtaposed with the technological advances of Western society, it is incomplete. This is because it does not take into account the efficacy of the reductionist method, which holds that the fewer the variables examined, the more reliable the data. In other words, this kind of knowledge acquisition depends on purposefully reducing reality to its various fragments. One fragment of reality can then be compared and analyzed with other fragments.

Nuu-chah-nulth peoples today, in concert with many other indigenous peoples, embrace the phrase that is variously translated as "everything is one," "everything is related," or "everything is connected." These phrases resonate deeply with an indigenous ethos and lived-experience, just as poetry and music may resonate deeply with some people. The Nuu-chah-nulth struggle towards wholeness meant a deliberate exclusion of any form of reductionism. In fact, any emphasis upon reducing the interconnected nature of reality was considered as a step towards weakness, which is why it was taught that everyone should learn to ask for help whenever help was needed.

If a cultural emphasis on the development and maintenance of relationships can be considered a strength, and if an emphasis on reductionism can also be considered a strength (at least in so far as it allows for the development of advanced technology), then this can add to Nuu-chah-nulth knowledge. My ten years of teaching First Nations studies has provided some confirmation that the foregoing statement has credence with indigenous students at Malaspina University-College (now Vancouver Island University). The students, in their papers and oral presentations, affirmed again and again that their understanding of their cultural heritages could be summarized by wholeness, interrelatedness, and interconnectedness. They recognized this as a strength, but they also recognized the strength of the scientific worldview. Therefore, I want to articulate indigenous strength and place it within a conceptual framework applicable to contemporary conditions. The term *haḥuułism* represents a synthesis of worldviews as it translates indigenous knowledge into a Western philosophical framework with the intention of suggesting the possibility of an equitable and harmonious working relationship between the two ways of life. As indicated in previous chapters, there are a number of thinkers today who are disillusioned with the current world order. Thomas Berry comments that "we are in trouble just now because we do not have a good story,"[2] and Fritjof Capra observes that "the major problems of our time ... are all different facets of ... a crisis of perception."[3] More to the point, Patricia Marchak and Jerry Franklin assert: "Perhaps the greatest challenges of all lie in the realm of the social contract."[4] By suggesting an updated approach to the way in which we manage the challenges of life, this chapter hopes to take one small step towards attempting to rectify the disillusionment with the current world order.

The *Qua* of *Haḥuułism*

Haḥuułism is a way of life based on an ancient view of an integrated reality, which consists of the unity of the physical and the non-physical. This ancient view of reality is supported by origin stories that were tested by an integrative method of research over a long period of time with consistent and reliable outcomes. This method of research is integrative because the practitioner is a physical being integrated with, and enlivened by, a spirit and a soul, which must work together in harmony in order to provide a sense of well-being. This method is the *ʔuusumč*, the vision quest. This general description of *haḥuułism* is *qua*, the nature of existence, the nature of reality, that which is.[5] *Qua* is a Nuu-chah-nulth word that includes every aspect of reality – empirical and non-empirical, physical and non-physical, normal and paranormal – which, in terms of human development, is reflected in the physical body and its non-physical soul/spirit.

Story as Theory

In the beginning each origin story, such as Son of Raven, Son of Mucus, Pitch Woman, Bear, and so on, was like a theory. This is not to say that ancient Nuu-chah-nulth peoples had theory, as such. Nevertheless, origin stories are similar to theory in that each one is testable by its own method, just as scientific theory is testable by its own methods. Another way in which origin stories are like theory is in outcome. When a theory has been tested over a long period of time and has reliable and consistent outcomes, the data derived from it become facts. The same is true for the testing of origin stories. The difference in outcome is that scientific data are usually confined to the empirical domain whereas *ʔuusumč* data are not.

When an origin story has proven valid as a guide to living, then it becomes an allegory, a dimension of truth. For example, the story of Pitch Woman can be considered as a proposition about the nature of good and evil. Pitch Woman behaves as though she believes that the way to live is by brute force, by coercion, without regard for law and order. She steals children, and for much of her life she is successful. Her beliefs are borne out by her successes.

The same can be said for the story of Bear, who stole salmon from a Nuu-chah-nulth fish trap, except that Bear, unlike Pitch Woman, was acting on a legitimate need to feed his community. The boundary that demarcates ownership of the fish trap is invisible, so, initially, Bear cannot see any markings of ownership. It is through lessons learned from these kinds of stories that ancient Nuu-chah-nulth sometimes considered empirical reality, in and of itself, to be unreliable because there were often hidden factors that lay beyond what could be observed by the human eye. It is not that empirical reality, in and of itself, is unreal or untrue; today, some scientists claim to understand about 5 percent of the physical universe.[6] What, to ancient Nuu-chah-nulth, proved unreliable about empirical reality was its interpretation, which was resolved through a systematic process illustrated in the story of Bear. Inherent polarity, presented as conflict, is resolved through a deliberate exercise of liberal democratic constitutionalism, here defined as the haḥuułic principles of recognition, consent, and respect, which enables life forms to continue in their diverse ways. Balance and harmony are forged out of inherent polarity.

The theory proposed in the story of Bear is that creation is set up in such a way that resources are juxtaposed with, rather than held in common by, any number of different communities. Resources that are in juxtaposition with different communities are in a natural state, and this is the way in which creation is presented. How these different communities respond to this juxtaposition of resources is a matter of choice. There is no obvious handbook to explain to these communities how to negotiate this juxtaposition of resources. Life forms have freedom of choice. In this theory, and in contemporary language, the choice for these different communities is between survival of the fittest or mutual respect, between brute force or sustainable living.

A simple example is fresh water, which, in the form of a river, is juxta-posed with the different communities that need it, such as trees, plants, insects, birds, bears, wolves, deer, and humans. A resource juxtaposed with a variety of communities presents a challenge regarding how to negotiate that particular condition of reality. Without a process of effective communi-cation between communities, there is no hope of sharing any resource in a mutually agreeable manner. Without any effective means of communica-tion, the default mode is survival of the fittest, whereby the stronger dom-inate and oppress the weaker. Mutual knowledge about one another is a simple necessity. The story of Bear proposes that it is possible to communi-cate between species in such a way as to develop a sustainable and shared protocol. The necessary cultural assumption is that, in order to communi-cate effectively between species, it is necessary to employ the integrative method that the Nuu-chah-nulth call ʔuusumč. During the time of the an-cient Nuu-chah-nulth, when there was no systematic large-scale effort to work cooperatively with both an indigenous method and a scientific meth-od, the only known way of ensuring effective inter-species communication was through the use of the ʔuusumč. Within this framework of reality it was unconstitutional for one life form to make decisions on behalf of other life forms. The Nuu-chah-nulth did not decide unilaterally how to relate to the Salmon people; rather, through a reciprocal form of communication the for-mer and the latter came to an agreement. In exchange for recognition by Nuu-chah-nulth peoples in the form of public ritual, the Salmon people gave themselves as food – the natural fulfilment of their purpose. Formal sharing is one response to the major challenge presented by a polarized reality in which resources are juxtaposed with competing communities of life forms.

The principal origin story for *haḥuuɫism* is Son of Raven and his com-munity. This is because it combines two major dimensions of a single real-ity – the physical and the non-physical. This story proposes that the primary characteristics of creation are:

1 Reality is one.
2 Reality is polarized.
3 Reality is composed of the physical and non-physical.

4 Non-physical reality is the source of physical reality.
5 Polarity is inherent and purposeful, and it necessitates:

a a negotiation of oppositions and the contradictions of apparent fragmentation;
b making choices between opposing realities;
c development of management regimes (such as governance of self, family, community, and the living environment); and
d multidimensional development of body, soul, and spirit.

6 Identity is like an insignificant leaf.
7 Light is both the source and goal of life.

The first three propositions are not unusual, and the fourth to seventh propositions may not be new. These are not meant to be presented as articles of faith but, rather, as propositions that can be submitted to research. Although I argue that, over millennia, Nuu-chah-nulth peoples have submitted their stories to the test of the *ʔuusumč,* with satisfactory outcomes, Western empirical research has been dismissive of our kind of knowledge system. This is well founded from the perspective of the scientific paradigm since not only do personal experiences vary widely in response to the same event but they also cannot be quantified (except in a very imprecise manner). My work is intended to contribute towards alleviating some of these scientific misgivings about personal experiences by suggesting that Nuu-chah-nulth origin stories, seen as theory, remain as testable today as they were in the beginning.

The Nuu-chah-nulth Test of Theory

By definition, a *ʔuusumč* is personal. *ʔuusumč* experiences often take place when a person is alone in the woods, by a prayer pool, by a river, or up a mountain. However, this does not mean that the data received are any less reliable than are the data received from scientific study. There are two important and compelling arguments in support of *ʔuusumč* as a reliable method for testing theory: (1) the *ʔuusumč* was considered necessary to survival (and indeed helped ancient Nuu-chah-nulth survive for several millennia); (2) among the Nuu-chah-nulth, the ʔuusumč was a universal practice employed consistently over time. Community-wide experiences and testimonies validated the notion that when the ʔuusumč was undertaken, the basic necessities of life – food, shelter, and clothing – were at least minimally met. And if the ʔuusumč was undertaken more effectively, the Nuu-chah-nulth experienced a corresponding increase in wealth of every kind, physical and non-physical.

The story of my great-grandfather Keesta is one example of this assertion about the relationship between the ʔuusumč and food on the table. Each family, like my own Atleo family, has a number of stories that relate the ʔuusumč to the provision of food. These family stories reflect the theoretical proposition of the story of Son of Raven and his community in that, when the appropriate ʔuusumč method was applied, success was assured, and, when it was not, success was not assured. When Son of Raven took an arrogant approach rather than a humble approach, he was always unsuccessful. The appropriate method includes both the proper ritual behaviour and the proper attitude. Consequently, whenever anyone had a poor hunting or fishing experience, there was immediate suspicion that either an inappropriate ʔuusumč method had been employed or that the ʔuusumč had been neglected completely. In the latter case, it was said the hunter or fisher must be wiišaʔ (one who does not properly prepare or who deliberately violates ʔuusumč protocols).

As mentioned in Chapter 4, Jeremy Narby was challenged to test his claims about the validity of indigenous knowledge, and he did so. As a consequence of his experience, Narby writes of Amazonian shamans that their "way of knowing is difficult for rationalists to grasp" and that "Western science has some difficulty with the possibility of both nonhuman intelligence and the subjective acquisition of objective knowledge."[7] Although the test conducted by Narby and his colleagues appeared to be successful, the results could not be considered reliable. In order for the scientific community to take Narby's findings seriously, the test would have had to involve a mathematically significant number of scientists or at least one of the three scientists involved would have had to possess the stature of an Albert Einstein. However, from the perspective of the Nuu-chah-nulth, Narby's findings are valid and confirm their own knowledge system.

Recently, I was at a conference with a relative of mine, Barney Williams Jr., who is from the neighbouring community of Tla-o-qui-aht. Barney, has much traditional knowledge and experience, has worked among First Nations as a counsellor. I sat with him at this conference and asked him: "How do you know that the ʔuusumč works?" I explained that the question was not posed from a Nuu-chah-nulth perspective, since we know that the ʔuusumč works, but from an empirically oriented perspective.

To illustrate my question, I told Barney the story of the traditional Beaver people who live in northeastern British Columbia.[8] The Beaver have a tradition of having strong dreams in which they make dream-kills. To ensure that a dream-kill is based in reality, the dreamer will, during the dream, mark the hoof of a cow moose. Then afterwards, the hunter will go out, find the dream trail, and follow and kill the moose. And there on the hoof he will find the exact mark that had been made during the dream-kill.

"Well," Barney said, "we did the same." He went on to say that, during the ʔuusumč, certain Nuu-chah-nulth whalers would have a vision of a whale hunt and would see a specific mark on the whale's fin. Then, after having gone out on the actual hunt, they would find the exact same mark on the captured whale. To strengthen the claim that visions can be directly related to reality, Barney then told the story of someone who wanted to be a whaler and went through the ʔuusumč process. When the prospective whaler emerged from his ʔuusumč, he was asked if he had seen the mark. "No" said the prospective whaler, but he insisted that he was ready, cleansed and purified. The whale hunt proved a disaster and his whaling canoe broke into pieces on the rocks.

The Beaver people employed a hunting strategy that demonstrated a rational connection between the hunter and the hunted through a medium of consciousness that enabled a link between two different planes of reality, the physical and the non-physical. When this strategy was employed, the results were the same as they are for any physics experiment whose purpose is to examine the nature of subatomic particles. The systematic connection between a dream and a successful hunt is similar to the systematic connections found by physicists between invisible particles. The same kind of data is produced by both types of methods on a consistent basis, which is one criterion for validation.

Examples of Nuu-chah-nulth Findings

But of what relevance is it to claim that the ʔuusumč method produces findings? Since the practice is personal and based in an oral society, in which detailed written records are not kept, can there be said to be findings? The answer may depend on the way in which reality is perceived and experienced. If the fundamental characteristics of reality are perceived to be dynamic and ever-changing, as in accord with ʔuusumč experiences, then it becomes problematic to keep records that are static. Moreover, how can empirical records accurately reflect non-empirical reality? How can the physical represent the non-physical? Many argue that the empirical and non-empirical are irreconcilable domains. In this worldview, static records will distort a dynamic reality. This dynamic reality was managed by Nuu-chah-nulth peoples through a method of storytelling that allowed for changes of detail from one telling to another (in the same way that reality presents itself as ever-changing).

At the same time, it would not be true to say that ancient Nuu-chah-nulth did not keep records. They did keep records – records that were allegorical, representative, and symbolic. These records are the stories, songs, dances, regalia, house posts, masks, headdresses, and so on that were instrumental in demonstrating powers that energized entire peoples along the Northwest Coast of British Columbia. This is one answer to the question about findings. For millennia, these findings contributed to community well-being.

Every time a hunter brought home the food, every time a hunter sang a new song, every time a ʔuusumč practitioner chanted a new prayer song, every time an artist presented a new mask or a new rattle, every time a dancer presented a new dance, every time a chief presented a new *tupati* (spiritual power) or new headdress, and every time a shaman performed a successful healing, ancient Nuu-chah-nulth recognized these as findings.

For the ancient Nuu-chah-nulth, the significance of a story is its contribution to community well-being. Thomas Berry might agree that this could be one definition of a good story: its ability to contribute to community well-being. The tale of Son of Raven and his community is one such story. At first the problem-solving strategies employed by Son of Raven and his community appear simply as a series of mistakes, a series of errors, but each time the story is told it concludes with success. To the ancient Nuu-chah-nulth, this kind of story reflects an ever-present truth, it reflects who and what we are and, therefore, always seems to take place in the moment. This is one explanation for the saying "our stories are true!" If the story reflects a credible truth about existence, then this truth is independent of time: it is timeless. Each listener is touched deep within her/his psyche by the timeless truth of the story. Each listener knows by experience the necessity of failure, which is then linked to personal development and eventual success. Each listener recognizes and accepts the mystery and partial knowledge assumed in the story. Ultimately, the story, like a parable, is about life as its truths are made manifest in the pragmatic provision of food, clothing, housing, and a measure of security of person.

Time and time again, explorers like Amerigo Vespucci, José Mariano Moziño, and Christopher Columbus were presented with the pragmatic results of indigenous ways of life based on story and the ʔuusumč method of knowledge acquisition. In March 1778, Captain Cook wrote in his journal of his encounter with the Nootka Indians that "they generally went through a singular ceremony."[9] What is described in Cook's journal is a typical (to Nuu-chah-nulth) sacred ceremony – one that continues to be practised today. To Cook's ears the sounds made are "halloaing,"[10] and to Vespucci's ears are a kind of "lamentation." These sounds were evidently prayer chants and, in the Nuu-chah-nulth case, always accompanied by the use of a rattle, the wearing of a mask or headdress, and the sprinkling of eagle down to signify a sacred peace. On each occasion of diplomacy between Nuu-chah-nulth leaders from different nations, prayer chants were offered and reciprocated. All of which is to say that these encounter-ceremonies were not haphazard events but symbolic articles of an ancient form of constitutionalism.

In a chapter of his book *Potlatch,* the late Nuu-chah-nulth artist, author, and actor George Clutesi provides an account of the same kind of ceremony described by Vespucci, Moziño, and Columbus,[11] except in this case it is presented by one who experienced it directly: "Often the king of each tribe would come in a flotilla of canoes. His own would be lashed to three or four of the largest ... with wide boards secured to the top to form a large surface. This served well ... for dances that might be planned by the visitors."[12] Then, while still at a distance, the flotilla "would strike up its own paddle song in complete unison with the stroke of their paddles" and, upon touching the white sand, an "imposing figure, dressed in all his most glorious regalia, gave forth with a lusty rendition of his incantation especially reserved for an occasion such as this."[13] Rather than describe the prayer song as a "lamentation" or as "halloaing," Clutesi describes it as "strong, clear, resonant, escaping the confinement of the cold waters to soar into the morning air and cut deeply into the dense forest behind the great lodges."[14] After this formal greeting by the invited guest, there is momentary silence until "a high pitched voice ... from the waiting throng" responds in kind. This reciprocity affirms that both sides are prepared, cleansed, and ready. Significantly and customarily, this "high pitched voice" is female, and the prayer song ends with "a cadence of proud dignified intonation."[15] This describes in the barest of detail extensive and elaborate ceremonies that incorporate and enact ancient constitutional principles of recognition, respect, and consent – principles that, to some degree, allow all to continue their own way of life in balance and harmony.

Given his upbringing and education, Captain Cook – in much the same manner as Chief Justice Allan McEachern of the British Columbia Supreme Court, after observing the symbolic articles of ancient constitutionalism presented by the appellants during *Delgamuukw* – did not recognize anything diplomatically meaningful in his encounter with the Nuu-chah-nulth. Neither Captain Cook in 1778 nor Chief Justice McEachern in 1997 had any balanced and reliable information about indigenous peoples that would allow them to see and understand a people whose lens on reality differed sharply from their own. For the ancient Nuu-chah-nulth, their way of life provided them with a lens through which to view their place in creation.

Origin stories provide the Nuu-chah-nulth with an orientation to life. In some ways, Son of Raven and his community are in the same position as first-year students who are experiencing a university environment for the first time. These students do not arrive at the university without credentials and an acceptable degree of skill level. It may be remembered that Son of Raven and his community began life with some power, some gifts, and some natural skills. They did not, according to Nuu-chah-nulth story, arrive at this university of existence without credentials, without some coping skills.

Transformation, of which only the smallest vestige remains today, was common to all the first people. They could transform into anything that they could imagine with the same ease that we today change our clothes and our hair colour. They were as familiar with the power of thought and the power of the word as we are familiar with technological communication systems. Have these early gifts, due to lack of belief and disuse, become extinct? Or might they just be dormant? That may be a very good research question. Nonetheless, given this start in life, Son of Raven and his community employ these gifts, these skills, collectively in their quest for the light that is kept in the box in the community of Ḱʷaaʔuuc.

Some Nuu-chah-nulth Beliefs Derived from Origin Stories

Beliefs about the nature of reality translate into principles, teachings, laws, and what today we would label policies. Principles may remain unchanged over time, whereas teachings, laws, and policies are adapted to changing socio-political circumstances. The unity of creation, the primacy of the non-physical over the physical, the dependence of the physical on the non-physical, the superiority of the insignificant-leaf approach to the swelled-head approach, the need to allow for multidimensional human development, the need to strive for balance and harmony, the need to develop protocols with all life forms, the need to respect all life forms, and, at the personal level, the need to be yourself – all these are principles that do not change.

In stories found throughout the indigenous traditional world, the antics of the Trickster – lusting, scheming, looking for an easy meal, the best deal, the fastest route, and the best advantage – reflect something that is common to human nature. They reflect basic human desires that are perfectly acceptable when expressed by babies but unacceptable when expressed by adults. One of the challenges of creation is to develop beyond this early stage of life. Fittingly, Son of Raven and his community act out this early phase of life-form development by attempting to take what appears to be the easy way to resolve a problem, which is just to go and take what they want without regard for rules, ethics, or protocols. This early phase of human development appears self-evident in babyhood but appears more and more inappropriate as a person grows older (although it remains undeveloped in some areas of the character of the soul or spirit). Ancient Nuu-chah-nulth would say of a person who, by behaviour or reputation, was unevenly developed that she or he was *wikiiš čaʔmiiḥta,* which means "[that person] is out of balance" or "is not in harmony." Son of Raven and his community began their developmental process in a state of immaturity, which is natural. Then they are led to discover the complex requirements of human development within a multidimensional reality. The story of Son of Raven and his community teaches that living requires the development of proper protocols that must be forged in an empirical context of apparent contradictions.

The suggestion found in the stories of Son of Raven, Son of Deer, Son of Mucus, Bear, and others is that the developmental process necessitated by the design of creation involves a complex union between the physical and non-physical, which requires a lifetime of struggle and effort to achieve. Although the story about Son of Raven as an archetypal representative of the purpose of existence has been told from the beginning, its lesson remains hidden to those who confine their observations to the empirical domain. Hence, this work, in its presentation of another research method – one that intends to integrate physical reality with non-physical reality – may add to and complement the established scientific method.

A Few Nuu-chah-nulth Teachings, or Laws, that Derive from Origin Stories

When I was a little boy my grandmother sometimes reminded me before I was sent out to play that I must not steal. On the surface it appears that the story of Son of Raven and his community is about stealing,[16] but its underlying meaning is about the proper way to grow up and earn a living. Son of Raven had to learn how to do things, how to negotiate the reality in which he found himself. He had to go through the accepted channels laid out in the original design of creation. He couldn't just take what he wanted. He had to learn respect for boundaries and borders. He had to acknowledge ownership, and then he had to start from the bottom, so to speak, when he was obliged to be born as a baby into the house from which he sought the light. Instead of immediately earning the right to take the light, Son of Raven first had to grow up and learn the ways of the house, the language, and the culture. Only then, after many years, was he finally allowed to have what he desired.

As a result, one Nuu-chah-nulth teaching is to work hard at whatever you find yourself doing in life. One translation of a Nuu-chah-nulth saying that admonishes hard work is: "Be full of life!" To this day, whenever I have a job to do, I can still hear elder family members of both sexes saying in Nuu-chah-nulth: "Be full of life!" You might notice that the need for *hard work* is assumed, whereas the need for *fullness in one's expression of life* is an admonishment or teaching. Juxtaposing "hard work" with "be full of life" is one way to define a path of life that is developmental.

Another teaching involves a saying that might be translated as "always feed your guests." So strong was the belief in the necessity of giving that it became institutionalized in the great formal feasts. The natural law, whereby giving is synonymous with life and not giving is synonymous with death, is illustrated in the story of Bear (told in Chapter 5). This law implies that, metaphorically, giving is associated with abundance, whereas not giving is associated with stagnation (since the circulation, or giving, of goods and assistance is essential to life).

It follows that we must help one another. The Nuu-chah-nulth teach that, if people don't ask for help when they need it, they are not kind. Asking for help when one needs help is to affirm the unity of creation; not asking for help when one needs help is contrary to wholeness. However, whenever a way of life is articulated by the constitutional principles of recognition, consent, respect, and continuity, this teaching applies not only to human beings but also to all life forms (as is argued in Chapters 4, 5, and 6). Each ancient household would apply its own perspectives, its own interpretations and family teachings, and to record all of them would fill a large library. These contemporary constitutional terms are indicated in Nuu-chah-nulth stories and I now describe them as *haḥuułism*.

Assumptions about the Nature of Existence

Haḥuułism assumes the oneness of reality in both its physical and non-physical aspects. The word itself is a metaphor for the unity of creation. When *haḥuułism* is divided into its parts, the syllable *ha* refers to a reality that is spatially nearby, and the syllable *ḥuu* refers to a reality that is spatially far away. In terms of the material universe, these two syllables, *ha* and *ḥuu,* when combined with the ending syllable *łi* to make the word *haḥuułi,* refer to a bounded territory with resources owned by a "wealthy one." The term *haḥuułi* can refer to the sovereign territories of a Nuu-chah-nulth chief, this being a model based on origin stories about K̓ʷaaʔuuc, Owner of All. *Haḥuułi* also applies to the non-physical domain, where *ha* refers to the physical, *ḥuu* refers to what is beyond the physical, and *łi* refers to the natural link between the two domains.

In descriptions of qualitative relations between life forms, the two syllables, *ha* and *ḥuu,* can imply either close relatedness or non-relatedness. One chooses which lens to look through. To the first people, to Son of Raven and his community, various parts of reality appeared to be fragmented. This empirical lens on their world seemed to indicate, just as it indicated to Sir Isaac Newton, that reality did not integrate *ha* (which is close by) and *ḥuu* (which is far away). It appeared that what was close by was not connected or related to what was far away. Consequently, it was natural for them to assume that the community from the spiritual domain was not related to the community from the physical domain. What did heaven have to do with earth?

In one sense then, these original people began life, in part, as empiricists. They were guided by their sight, something that parallels an early phase of human development. Kindergarten training begins by focusing on the concrete before moving to the abstract. Consequently, when Son of Raven and his community began their quest for the light they assumed a (physically distant) *ḥuu* relationship to the owner of the light, the Creator. *Ḥuu,* in this case, assumes that reality is fragmented. Psychotherapists may call this fragmented condition dissociation, or anomie, which is considered dysfunctional. In the end, Son of Raven discovered that his relationship to the Wolf community was not *ḥuu* (distant) but *ha* (close), so closely related, in fact, as to be *astonishingly intimate.* This early discovery about the nature of reality is the foundation upon which *haḥuułism* is built.

An important assumption embedded within Nuu-chah-nulth language concerns time. The expressions, *qʷaasasa sqʷi, qʷaasasa iš,* and *qʷaasasa uλ,* which, respectively, translate into English as "that's the way it was," "that's the way it is," and "that's the way it will be," are common Nuu-chah-nulth expressions. They integrate and hold in balance the apparent tension between linear time and the eternal moment. The ending of each word represents the linear past *(sqʷi)*, present *(iš)*, and future *(uλ)*, but the root of each word, and consequently the source of each, is *qʷa* – that which is – the present moment. These three common Nuu-chah-nulth expressions speak of linear time contained in the present moment – that which is. Past, present, and future are *tsawalk* – one. It may be that the greatest knowledge (at least according to those who became ?uuštaqyu [shamans], which is an accomplishment whereby, through a lifetime of struggle, the body, soul, and spirit become an integrated whole) about the nature of reality is expressed in the wonder of the unknowable and the elegance (in the scientific usage of this term) of the mystery of reality, which has to do with the passage of time being in balance with the present moment.

To Son of Raven and his community, language at first reflected an early phase of development, as seen in the everyday usage of the words *ha* and *ḥuu*. As a consequence of the revelations gained from practising the ?uusumč method to investigate the nature of reality, however, the two words were brought together to form *haḥuułi,* which I refer to as *haḥuułism* – the integration of the empirical with the non-empirical, the physical with the non-physical. Then it was that all ancient Nuu-chah-nulth came to understand the integrative name Ḱʷaa?uuc, owner of that which is far and that which is near, owner of both the non-physical domain and the physical domain. Language is *qua*.

Human Development Issues

From a Nuu-chah-nulth perspective, it may be futile to attempt to explain human development issues within the context of a creation that is mostly mysterious because to do so would imply knowledge of ultimate purposes. In spite of this limitation, however, some things are known about human development. Identity is one. Personal boundaries are necessary to a healthy identity. Megalomania is an issue of human development because its actions imply that a personal boundary has been inflated. Son of Raven illustrates this condition when he transforms himself into a giant king salmon, unlike others who, by former agreement, transform into the much smaller sockeye salmon. In the end we learn that the actual personal boundary of Son of Raven, in comparison to the owner of the light, the Head Wolf, is like that of an insignificant leaf. In other words, Son of Raven, in his journey through life, had to discover his identity in relationship to creation. What is this identity? From the perspective of this origin story, it is very small, like a leaf that is so tiny it could be swallowed without difficulty. At the same time, this natural identity becomes the key to unlocking some of the mystery of creation. A realization of this insignificant identity in the context of creation proves to reveal an intimate unity with the light.

In balance with this relative insignificance of being is that characteristic observed by Vespucci when he wrote that "everyone is lord of himself."[17] "Lordship of self" is a phrase that reflects an ancient teaching found in the everyday expression *qʷaasasa iš,* literally, "that is just the way he or she is," which implies that the way of a person is a natural characteristic of her or his being. From an ancient Nuu-chah-nulth perspective, the lordship of self is taken for granted, it is taken as a necessity in the context of the perpetual mystery of polarity, where *qʷaasasa iš* applies equally to good people and bad people, creative people and destructive people. One can be lord of oneself on either side of the polarity of existence.

Consequently, in order to secure a prevailing sense of well-being towards community, the lordship of self must predominantly be on the creative side of polarity. In this sense of lordship of self there is a parallel to the contemporary notion of self-reliance, the development of which, in Nuu-chah-nulth society, meant forging a complex of characteristics within both the soul and the spirit. As a generalized value, the assumption that each person is lord of him- or herself, combined with the practice of asking for help when help is needed, permitted mutual recognition of, and mutual respect for, the integrity of each life form in the context of community.

Although *haḥuułism* describes a way of life that is inclusive of everything, and although it is characterized by interconnectedness, it is also defined by a necessary struggle for balance and harmony. This may be considered a contradiction in empirical terms because the notion of interconnectedness is inconsistent with balance and harmony. Reality does not appear to be inclusive, nor does it appear to be interconnected. In fact, as

Charles Darwin observed, reality presents itself as a ubiquitous struggle for life governed by a process that selects the fittest for survival. How can one balance and harmonize a disconnected reality? Son of Raven and his community made the same assumption, which is why they first thought to simply take what they needed without regard for law and order. However, when this same reality is subjected to the integrated research method of the ʔuusumč, then another kind of process and purpose to life can be unveiled.

Haḥuułism can be defined by the struggle for balance and harmony, as is suggested by the original title of this book, Protocols of Tsawalk. "Protocols" refers to agreements or treaties between life forms that must compete for resources on one planet, known to Nuu-chah-nulth as Haw'iłume. The phrase "compete for resources" has both a negative and a positive interpretation, and it is the latter that haḥuułism seeks to address.

What this means for the human development process (which is still emergent, still mostly undefined, but guided by the metaphor of light) is a system of life management to which I refer as "protocols of tsawalk." These are conventions that exist between life forms and that move competitive relationships away from conflict and towards harmony until all the constitutional principles of life – mutual recognition, mutual consent, and mutual respect – allow for the continuity of all life forms.

Protocols are hammered out of apparent contradictions over an appropriate period of time. For example, in the scientific story, life forms emerged billions of years after the Big Bang. This period of time is necessary because the fact of life must be explained according to a theoretical beginning based on contemporary "evidence." Empirically, life and time are irrefutable, and billions of years appear to be an appropriate period of time within which to imagine and theorize a gradual, evolutionary emergence of life. However, from a Nuu-chah-nulth perspective, although time is also factual in the empirical sense, it becomes non-existent, or non-significant, through the experiences of the ʔuusumč. The Big Bang story is one way to describe beginnings. It has empirical validity. What happened after the beginning,[18] according to the ancient Nuu-chah-nulth, is a matter of story and the natural tendency of stories to have multiple interpretations.

This issue of multiple interpretations regarding the nature of reality is illustrated in the story of Ałmaquuʔas, Pitch Woman, who stole children so that she could have a family. She achieved her purpose, and so her interpretation of the nature of reality proved successful and empirically sound, albeit only for a time. In other words, both sides of an empirically observed polarity, the good and the bad, can be verified by personal experience. Perhaps because she was overwhelmingly powerful and could dominate and take advantage of others, Pitch Woman accepted that hers was the primary version of reality. She would agree with the theory of survival of the fittest. This story parallels the recent story of colonialism, in which dominance and advantage was rationalized by the logic of evolutionary theory, which prevailed over law and order, at least in so far as these applied to indigenous peoples.

What continues to prevail today in terms of dominance and advantage, in both mainstream scholarship and mainstream media, is the notion that science is the only source of authentic knowledge. Of course, what is prevalent in mainstream scholarship is also prevalent in the school system, and what is prevalent in the school system is reflected in any given society's worldview. There have always been alternative worldviews but only at the margins of power and influence. With respect to indigenous knowledge, for example, a leading educator recently remarked: "There is nothing there."[19] This statement may be sincere, but it is made from a recent historical context – one that completely excludes indigenous knowledge systems from any part of its school curricula. In any case, if "there is nothing there," then the colonial agenda should have been more thorough. It was not. So now we are experiencing a gradual reversal of colonially based views about Aboriginality. On 11 June 2008, the leaders of every political party apologized to the survivors of the residential school system, which was intended to eliminate indigenous lifeways. If, as I argue, there is something in the lifeways of indigenous peoples, then this "something" may fly in the face of early scientific beliefs. How did Aboriginal lifeways withstand and survive the test of time? Why did they not disappear from the earth?

Purpose of Indigenous Lifeways?

The Nuu-chah-nulth word for "completed person" may also be translated as "shaman." There are many kinds of completed persons, just as there are

many accomplishments in many different fields. The major difference between the training of ancient Nuu-chah-nulth "completed people" and the
formal training of accomplished contemporary people is that the former
integrated the physical and non-physical aspects of the person whereas the
latter does not.

Moreover, since the Nuu-chah-nulth view of reality is never fixed, frozen, or objectified, the phrase "completed person" must be understood
within the context of a dynamic reality. When the conditions of reality
present impossibly traumatic and destructive forces, a completed person
may have a broken spirit and a humble heart. In the story of Pitch Woman,
the devastated but completed person is the wife of the chief. The suggestion
here is that traumatic occurrences affect the earth community and the community of K̓ʷaaʔuuc equally. This is because no person could be completed
without the cooperation of powers from the spiritual domain, which means
that a completed person must have a direct connection to the Owner of
Reality. Therefore, as is illustrated by Son of Raven when he was born into
the house of K̓ʷaaʔuuc, a broken spirit and humble heart allowed K̓ʷaaʔuuc
to connect with the person so afflicted. Under conditions of extreme trauma,
a broken spirit and humble heart can be defined as a healthy identity – one
that is in a good relationship with K̓ʷaaʔuuc.

Protocols and agreements are not givens but must be mutually developed
and negotiated. This is a simplification of a difficult, complex, and ongoing
process. Choices create the need to acknowledge boundaries. Without the
acknowledgment of the integrity of boundaries conflict becomes endemic.
The notion of boundary integrity is similar to the recent scientific idea of a
holon. Each whole is a holon. Elisabet Sahtouris gives credit to Arthur
Koestler for this idea: "The philosopher scientist Arthur Koestler suggested
we call each whole thing in nature a *holon* – a whole made of its own parts,
yet itself part of a larger whole. A universe of such holons within holons is,
then a *holarchy* – in Greek, a source of wholes."[20] Wholeness and holons
have boundaries – national, institutional, personal, biological, or environmental – and appear to be a natural principle of creation. In the same way
as the empirical universe is now perceived to be made up of bounded areas,
so, too, is the nature of reality suggested by the story of Son of Raven and
his community.

The Constitutional Metaphor of the *Huupak*ʷ*an'um*

In the story of Son of Raven and his community, we find that the Creator kept the light in a box-like structure known as a *Huupak*ʷ*an'um*. This light is the source of all life and all knowledge regarding how to live not only in relation to each other and the rest of creation but also in relation to the Creator. As a result of exposure to, and apprehension of, this light, one recognizes that the *Huupak*ʷ*an'um* has at least four constitutional principles. The first three are recognition, consent, and continuity, and the fourth is *iisʔaƙ*, a form of sacred respect. To practise *iisʔaƙ* means to hold all creation in reverence so as to acknowledge, recognize, and affirm the Creator. Accordingly, in the context of human development, neither an attitude of respect nor the practice of respect comes naturally. Both the attitude and practice must be learned, worked at, and then maintained with continued effort and persistence, often in the face of trying, disrespectful circumstances, which are guaranteed by the polarized nature of existence.

Ancient Nuu-chah-nulth government is modelled after the pattern we observe in the Wolf community. The head Wolf is Ḱʷaaʔuuc, the Creator. Ḱʷaaʔuuc is the *hawił*, chief, and Ḱʷaaʔuuc's articles of constitutionalism are kept in a *Huupak*ʷ*an'um*. Since, in principle, this leadership position is permanent, the only way to replicate this permanence on earth is to make it hereditary, either by the letter or by the spirit. The letter of inheritance is biological, whereas the spirit of inheritance involves adherence to the principles of Ḱʷaaʔuuc. The eldest son inherits the role of chief through biological descent, unless he is found wanting in the qualities necessary for leadership. If required, the chieftainship is passed to the nearest relative who is deemed, by a council of elders, to be qualified.[21] All these characteristics of governance and leadership are reflected in the ancient ways of the Nuu-chah-nulth.

Hidden in the story of Son of Raven is the great teaching, or law, of *yaaʔakmis*. *Yaaʔakmis* is the Nuu-chah-nulth word for "love," and it is also the word for "pain." It is the perfect word to reflect the polarity of existence and its potential unity – love on one side and pain on the other, with both merged into one experience. Any oral or written attempt to describe this would be inadequate and clumsy, for it is a truth that includes but necessarily transcends cognitive apprehension. It may be the critical link between the physical and non-physical domains. All Nuu-chah-nulth chiefs are taught to love their *musčim,* their people. This kind of love involves integrating love and pain, both of which exist naturally and independently within a polarized reality. Undeveloped, unmanaged, ungoverned, or unexamined love and pain remain separate and incomplete. For the ancient Nuu-chah-nulth, pain as the co-root of love is part of the elegance of the mystery of creation. The Head Wolf – or K̓ʷaaʔuuc, the Creator – loves his creation in the context of developmental challenges, which renders pain unavoidable. This quality of leadership, not always adhered to, is what is taught to Nuu-chah-nulth chiefs.

Balance of Power

In a hierarchical system of governance, the greatest danger is the potential for dictatorial abuse. An aspect of ancient forms of traditional governance observed by Vespucci and others was both the absence of institutions of punishment (e.g., jails) and the absence of totalitarian rule. This fact has always been dismissed on the grounds that indigenous peoples were thought to be primitive and to belong to communities too small to have evolved into advanced societies. However, from a Nuu-chah-nulth point of view, part of the reason for this absence of jails and totalitarian rule is *iisʔak̓,* sacred respect for all life forms.

Iisʔak̓ can have two applications: (1) respect for other life forms and (2) respect for oneself. Confusion results when respect for self conflicts with respect for others. In practice, respect translates into recognition that the self is part of a greater whole and that this means that it must be considered a part of the self of each plant, animal, and life form within the natural environment. In addition, each self has a distinct way of life expressed by the common Nuu-chah-nulth saying *qʷaasasa iš* – that's just the way that person (or it) is. Each life form has a natural path through life that depends upon the choices that she/he/it makes. The choices of Son of Mucus and Pitch Woman illustrate this point. Son of Mucus shoots arrows towards his heavenly abode to indicate his natural path through life, while Pitch Woman, in her desire for community, steals children, which creates a life path that leads neither to children nor to family.

Some Possible Translations of the Principles of *Hahuulism* for the Future

Hahuulism, as a way of life, does not promise utopia but, rather, a struggle fraught with dangers and challenges. Sometimes these dangers and challenges can be overcome and resolved, and sometimes they cannot. The latter condition seems to prevail for our planet today, although there are still some who have faith in the power of science to adequately address current problems.

Translating the principles of *hahuulism* within a contemporary setting faces natural and difficult challenges. The first challenge is that liberal democracy is still in the early stages of development. That it can now lay claim to being a global phenomenon is not necessarily hopeful news because a global event means a global-scale polarity. Regarding the emergent state of democracy, John Dunn comments: "If there really could be [any hope for democracy], what is quite clear is that we are not for the present moving towards it. Until we do, we should at least expect to go on paying the price for the scale of our failure to do so."[22]

The second difficult challenge is the fact that *hahuulism* represents a dramatic shift in worldview. Nuu-chah-nulth stories suggests that a characteristic of life forms is that they are naturally resistant to change. When Son of Mucus returned to earth to transform the first peoples into the biodiversity that exists today, he met with fierce resistance. Yet, inevitably and purposefully, the very means of resistance became part of the means of transformation. It is recognized that a shift in worldview is a complex process that, even under ideal conditions, requires many generations to accomplish; under less than ideal conditions, such as extreme climate change, which suggests catastrophic outcomes, change may come suddenly. A worldview has a natural lifespan that may be considered in terms of maturation. At maturity, a worldview may culminate in what is known as "fullness of time." I discuss the concept of "fullness of time" in *Tsawalk: A*

Nuu-chah-nulth Worldview, in which I indicate that a society's trajectory fulfills a stage of its existence.[23] This trajectory can tend towards stability or instability, balance or imbalance, harmony or disharmony. A trajectory achieves its "fullness of time," or point of maturation, when it fulfills a natural law of completeness. This completeness is either positive or negative and can be illustrated, at the micro level of analysis, by people who have drug and alcohol addictions. A negative point of maturation is referred to as "hitting bottom." A society or nation or planet can hit bottom in the same way as can a person.

Given the current trajectory of the globe indicated by climate change, it appears that the earth may be close to its fullness of time. If this is true, then a shift from one worldview to another may suddenly become possible. It is this possible shift that may allow some of the principles of *haḥuulism* to be considered as *part* of a set of possible solutions to a global crisis. Any attempt to apply the principles of *haḥuulism* to society must necessarily be accompanied by errors. That is one of the first teachings of the story of Son of Raven. The second and equally important teaching is how to apply the integrative method to human development (which includes body, soul, and spirit). Exactly how this integrative method is defined must always be left open because ultimate purposes remain, due to the infinite nature of reality, outside of human apprehension. Nonetheless, many of the issues that currently plague humankind may begin to be addressed by a general application of the principles of *haḥuulism.*

Good Government, Rule of Law, and Civilization

Governance, like any form of management regime, involves a developmental process. Good government contributes to community well-being while bad government does not. Good government employs the principles of *recognition, consent,* and *respect* with regard to all life forms so that each can *continue* its way of life. These principles form the rule of law and create civilization. The inherent polarity of reality means that, even when good government, rule of law, and civilization are in some measure achieved, the maintenance of these ways of life requires the same kind of effort and hard work that it takes to create them. The history of the rise and fall of civilizations and empires attests to the enormous challenges and difficulties involved in the maintenance of any way of life.

Education and Training

The colonial education process, which attempted to remake indigenous peoples in the image of Europeans, violated the principle to begin educating children from where they are – developmentally, linguistically, and culturally. This violation guaranteed educational failure. However, under a new worldview, it will not be enough to teach children from where they are because the current educational process does not develop the whole person. Teaching must begin not only from where children are but also from "what" children are. This principle necessitates an integrative training program that develops the body, mind, and spirit. Compulsory public education promotes cognitive development but ignores spiritual development.

Hahuułism's principle of educating and developing "what" the child is involves an awareness that spiritual growth must accompany physical and intellectual growth in such a way as to strengthen each area without compromising any other area. This new educational system will require a major shift from a focus upon material reality to a focus upon the different dimensions of human experience, which, to date, have not been generally accepted as valid areas of study. Indigenous knowledge systems can make major contributions to helping us to make this enormous and difficult shift in worldview.

The world is full of stories; however, as I note several times throughout this work, one story prevails. Thomas Berry argues that this is not a good story because it has *no inherent meaning* and, therefore, is indifferent to the well-being of human societies. This story does not reflect the human struggle for meaning, for good government, for the rule of law. It does not reflect the meaning of a social contract that even self-organizing life forms such as microbes and bacteria seem to have learned, which suggests an alternative kind of story, one with purpose and meaning.

If this primary story needs to be changed, how can we do this? Can the story be changed by argumentation? Can the story be changed by a scholarly elite? Can the story be changed by the example set by the most

powerful nation on earth? Perhaps one answer may be found in the story
of Son of Raven and in the story of Bear. In both these stories change is
brought about by personal experiences – not by argumentation, not by the
presentation of grand ideas, not by a scholarly elite. In the story of Bear, the
solution to the problem of competing·interests over the salmon is a result
not of argument but of experience. The Bear brought the Nuu-chah-nulth
person to his home to demonstrate the need for salmon and then applied
reasoned argument in forging an equitable agreement to share resources.
This story illustrates that personal experience is an effective element in
proving a point, but it is only a beginning: to it must be added reason and
logic. The same is true with regard to the story of Son of Raven, who came
to be born into the household of the Owner of the Light.

In this new educational system, personal identity will include current
knowledge about healthy human growth, but this will be balanced by the
metaphor of the *insignificant leaf,* which reminds us that we live in a universe that remains mostly mysterious, even to our most advanced scientists.
This insignificant leaf, the humble identity taken on by Son of Raven, may
be considered the archetype of an attitude that, in ancient societies, prevented the rise of autocratic rule while allowing for a keener perception of
the nature of reality (which, of course, implies an effective means of
education).

Remembrance

A central ceremony of *haḥuułism* involves periodically, publicly, and rever-
ently acknowledging that humans are characterized by short-term memory.
Humans have a tendency to forget; they are easily distracted. Humans have
a tendency to prefer the "quick fix." Humans have a strong streak of Son of
Raven and Coyote, both of whom are a lot of fun and often hilarious.
Seasons of relative peace and prosperity bring on a sense of false security.
This is the time of decay, when people begin to take the easy way through
life, when the ego assumes superiority and the human identity becomes lost
in the contradictions of polarity, when nations and empires begin to fall,
when people lose their way and forget their teachings. The ancient Nuu-
chah-nulth guarded against falling into such times with a periodic remembrance ceremony called a *ƛuukʷaana,* which means "we remember reality."

We remember who we are, perched precariously in a balance between creation and destruction, between the primeval darkness and the promise of the light to which Son of Raven became heir. This ceremony addresses the problem of collective memory. As of this writing, many nations that dismiss indigenous participation in their histories do not have a collective memory. If it is true that humans have a tendency to forget, then a new educational and socio-political system might be developed in concert so as to collectively address this problem either through some mutually agreeable formal institutional means or through some equally effective informal personal means.

Homelessness

"Everything is one" is one of the principles of haḥuułism taught by the story of Son of Raven and his community. Initially, Son of Raven thought that separateness-of-being was part of the natural order, but then he discovered his community's intimate relationship with the Creator. From that time onward, relationality became a first principle of ancient constitutionalism, and it was ritually expressed as recognition. The Nuu-chah-nulth practice of hamipšiƛ involves hosting a feast and, according to established protocol, publicly recognizing a visitor with a gift. The word itself means "to recognize." A visitor is recognized either by blood or by a legal relationship, which is one example of an underlying principle that became generalized between all life forms. Relationality and recognition make sense both in the context of Darwin's theory of common descent and in the context of the theory of quantum mechanics, according to which the universe is one giant energy field. Here, the phrase "common descent" refers to the fact that all life forms are derived from the same matter or from a singular source of energy.

Son of Mucus set the standard for high-quality relationships between all life forms. In terms of policy, these standards encompass social acts that provide for the basic needs: food, shelter, and clothing. In this kind of state, homelessness would be unconstitutional because homelessness is non-relational, a state of fragmentation. Non-relationality and fragmentation describe the negative side of polarity. Haḥuułism seeks to integrate life into wholeness.

Poverty

Sharing and generosity were so important to ancient Nuu-chah-nulth societies that to share meant to live and not to share meant to die. After contact, however, the potlatch system eventually succumbed to abuse until it lost its original intent, which was to maintain balance and harmony between and among all life forms. Prior to the loss of the purpose of the potlatch system, reciprocity was common between the Nuu-chah-nulth and the salmon, between the Nuu-chah-nulth and the cedar tree, between the Nuu-chah-nulth and the bear, and between my great-grandfather Keesta and the whale. Reciprocity also prevailed between the wolf and the deer and between Ḱʷaaʔuuc and Son of Raven. Thus, reciprocal gift giving and assisting one another became a law that, when properly followed, meant that poverty was non-existent. This may be why, when the Montagnais heard of widespread poverty in Europe, they were astounded and thought that it would not take much intelligence to remedy the situation.

Maximum Exploitation of Resources

It may be argued that climate change has been caused by environmental imbalances created by a free-enterprise system that, for the past five hundred years, has practised the philosophy of *maximum exploitation for maximum profit.* All the primary-resource industries, including oil, gas, coal, fish, and forestry products, have practised this philosophy to such a degree that imbalances in the environment and shortages of resources have become global issues. Under *haḥuułism,* this practice of disharmony and imbalance would be unconstitutional. The origin story about Bear, discussed in Chapter 5, illustrates the need to share resources held in common so that all may continue to exist as one whole. Sharing is a natural law that requires the development of protocols of mutual understanding, recognition, consent, and respect.

Corporations

The formation of unions for various purposes seems to have occurred since the beginning of time. Certainly, in Nuu-chah-nulth stories, the promotion of family and community is clearly intentional. Ḱʷaaʔuuc, the Creator, sends Son of Mucus down to the earth to restore family and community to the people of Ahous after Pitch Woman steals their children. *Haḥuuḷism* teaches that the purpose of any union between and among people is to enhance the well-being of family and community. Corporations, under their current and only constitutional mandate, which is to make money, are in violation of natural *haḥuuḷic* law. For corporations, the creation of wealth has become a purpose in and of itself rather than a fulfilment of *haḥuuḷic* law's requirement to provide for the well-being of family and community, which includes all life forms on planet earth.

Conflict

Nuu-chah-nulth origin stories indicate that creation began in conflict. Son of Raven and his earthly community took an oppositional approach to the Wolf community, as did Bear towards the Nuu-chah-nulth community. Pitch Woman sought to destroy family and community. The scientific story of creation also began with an unimaginable act of destruction, known as the Big Bang. According to Nuu-chah-nulth and Western scientific perspectives polarization is inherent to reality. If this is true, polarity ensures the co-existence of contrary forces; consequently, the potential for conflict is always present.

The *haḥuuḷic* approach to conflict is to accept reality as inherently polarized and then to seek ways to manage it. So long as reality is inherently polarized it follows that, when conflict is resolved through mutual agreement, this agreement will last only so long as it is mutually enforced. As soon as one party defaults on the agreement the conflict resumes. In this sense, the creation of agreements involves a co-scripting of reality. Both parties come together to write a script that they mutually agree will be the basis of their cooperation.

Management of polarity is one principle of *haḥuuḷism* that applies in both micro and macro worlds. Genetic studies show the natural conflict that takes place between the human immune system and invasive parasitic micro-organisms. Because bacteria are naturally present in the human body, disease is always lurking should the immune system weaken or break down. Yet, the simultaneous existence of bacteria within a well-balanced immune system is necessary to health. This is one example of the success-ful integration of polarity. Managing polarity includes recognizing that human development involves the need to face challenges. These challenges may be interpreted as conflicts that produce negative results or as challen-ges that are necessary to maturing towards wholeness, towards integration of being, towards becoming a completed person.

Assumptions

One possible subtitle of this work is *Assumptions about Reality.* Assumptions are identical to beliefs. What is assumed, or believed, is that existence has inherent characteristics that are not obvious and therefore might not be dis-covered. Assumptions, like non-physical reality, are hidden and form the foundation upon which perspectives, laws, policies, and practices are situ-ated. An assumption about the nature of reality is exemplified in the phrase "Favoured Races,"[24] which features in the subtitle of Darwin's major work. The identity of the "favoured races" and the identity of those not favoured translates into the coercive and hegemonic rule of the colonizer over the colonized. The St. Louis World's Fair, the US Declaration of Independence, the Canadian Indian Act, the frontier motto "the only good Indian is a dead Indian," the long-standing Canadian Indian education policy (cultural genocide, residential schools, the constitutional ban on sacred Aboriginal ceremonies), all ensured that Aboriginal societies would enter the twenty-first century as front-runners in every known human dysfunction, from educational failure to socially maladaptive behaviour – addiction, suicide, family violence, foetal alcohol syndrome, infant death syndrome – a variety of diseases, and incarceration. My hope is that the application of the prin-ciples of *haḥuułism,* which postulate the unity of creation through the de-velopment of certain protocols, may lead to the continuity of the lifeways of all life forms.

The Light

In the story of Son of Raven the first people begin life in darkness. The darkness is a metaphor for ignorance, for lack of knowledge, and for being unclear on the nature of reality, which creates an inability to function effectively. The darkness precedes the light and appears natural and necessary. Reality for created beings does not begin with an arrived-completeness, or maturity, but, rather, with that which precedes the light – darkness. However, Son of Raven and his community were not left helpless in the dark but, rather, were provided with enough personal powers and gifts, and just enough information, to enable them to add to their knowledge base and to make the necessary discoveries. This process also served to help them to grow and to develop into better kinds of beings. Similarly, today's peoples are not helpless: a polarized reality always guarantees its opposite – in this case, help.

This book represents a Nuu-chah-nulth perspective on this developmental process, which must begin with darkness and move towards light. It is an emergent perspective that requires the addition of other perspectives in order to be more complete. In the beginning, darkness restricts perception in terms of ideology, philosophy, and worldview. However, over time, as indicated by such stories as Son of Raven, Eagle, Bear, Deer, and the Salmon people, the gradual influence of light (a metaphor for seeing beyond the physical) allows for the realization that what is beyond the apparent confusion and fragmentation of a physical universe is a unified diversity of being. The ancient Nuu-chah-nulth began to address the reality of this diversity of being by applying the constitutional principles of recognition, consent, and continuity, which are founded upon a sacred respect for all life forms. The gradual realization of this unified diversity of being does not happen without challenge, struggle, hard work, and passion, all of which appear to be necessary to healthy growth and development at all levels – personal, familial, communal, national, and global.

The current global crisis, represented by climate change, a looming energy crisis, rampant diseases, the possibility of nuclear war, terrorism, and related problems, replicates, in principle, the original conditions of creation in which Son of Raven and his community found themselves. Metaphorically, the current global crisis is the darkness that precedes the light. According to the ancient Nuu-chah-nulth, light is represented by the principles of recognition, consent, respect, and continuity. And light is the final destination of life.

Does the story of Son of Raven and his community provide a credible theory regarding how to acquire the light needed to meet the challenges of today's global crisis? Is this crisis a metaphor for an original darkness that has, in principle, the same kind of solution as that found by Son of Raven?

Notes

Introduction

1 Gilbert M. Sproat, *The Nootka: Scenes and Studies of Savage Life* (1868; reprint, Victoria: Sono Nis Press, 1987), 3-4.

2 Evalyn Gates, *Einstein's Telescope: The Hunt for Dark Matter and Dark Energy in the Universe* (New York: W.W. Norton and Company, 2009). Gates clearly shows the uncertain nature of the data from the most recent scientific findings about the nature of the universe (see also note 3 below). Since there appears to be no scientific consensus about the nature of the universe, this suggests that any conclusions about climate change must include the possibility that it is affected not only by human activity but also by cosmic activity.

3 Gates, in *Einstein's Telescope,* reveals startling new evidence about the nature of the universe. First: "Since every bit of mass in the Universe is attracted to all of the other bits of mass in it, their mutual gravitational attraction should be putting the brakes on the outward rush of the galaxies. But instead of slowing down, the supernova teams found that the Universe appears to be speeding up – our Universe is accelerating, flagrantly contradicting all previous predictions and implying the existence of a substance that challenges our most basic ideas of physics on both the largest and the smallest scales" (29). There is no way to overstate the "challenges" presented to the scientific community. First of all, the simple fact of acceleration seems to eliminate the long-standing scientific notion of a "Big Bang." In the context of a universe in which "every bit of mass ... is attracted to all of the other bits of mass," which has a braking effect on any explosive motion, there is more to the nature of the universe than can be currently explained by science. Second: "Something has to be fueling this acceleration ... Something that has been termed *dark energy.* Dark energy is the major component of the Universe, constituting 72% of everything that is. We don't know what dark energy is, and we don't have any outstanding candidates either" (29).

4 Quoted in H.J. Vallery, "A History of Indian Education in Canada" (MA thesis, Kingston, ON: Queen's University, 1942), 114.

5 Basil Davidson, *The Lost Cities of Africa* (Boston: Atlantic, Little, Brown, 1991). For example, regarding some early African states, Davidson comments: "They are all civilizations of the savannah, of space-defying plains of grass and great seasonal heat; and

they are all built on urban trade and a pastoral-agricultural economy. In all of them the
great rivers of West Africa exercise a formative influence. The earliest is Ghana, already
a centralized state when the Arabs first mention it in A.D. 800. The second is Mali,
taking its rise in the thirteenth century but persisting until the seventeenth. A third is
Kanem, which later became Bornu; and a fourth is Songhay, whose power and prestige
would cover the fifteenth and sixteenth centuries ... Some of these states were the con-
temporaries of early medieval Europe, and may at times be accounted superior to it in
civilization" (79).

6 Charles C. Mann, *1491: New Revelations of the Americas before Columbus* (New York:
Alfred A. Knopf, 2005), 177.

7 Ibid., 12.

8 Amerigo Vespucci, *The First Four Voyages of Amerigo Vespucci*, reprinted in facsimile
and translated from the rare edition (Florence, 1505-06) (London: Bernard Quaritch,
1893), in the library of Dr. Andrew Taylor. See also C. Edwards Lester and Andrew Foster,
*Life and Voyages of Americus Vespucius with Illustrations Concerning the Navigator,
and the Discovery of the New World,* 4th ed. (New Haven, CT: Horace Mansfield, 1851).

9 Mann, *1491,* 13.

10 The quotation is taken from Chinua Achebe's *Things Fall Apart: A Casebook,* ed. Isidore
Okpewho (New York: Oxford University Press, 2003), 3. There are few works outside
of the Western canon that have received anything close to the attention given to *Things
Fall Apart,* and for good reason. In its home continent of Africa, it is arguably the most
widely read book, next to the Bible and the Quran. Outside of Africa, it has become part
of a global literary canon; it exists in close to sixty languages (including English), with
total sales of nearly 9 million copies since its publication in 1958, and it is included in
some of the world's most prestigious literary series. It would be hard to find many institutions of higher learning in the English-speaking world that have not listed this novel
in courses in the humanities and social sciences, at both undergraduate and graduate
levels. It is also a highly favoured text at the secondary school level. It occupies such an
important place in critical and cultural discourse because it inaugurated inquiry into the
problematic relations between the West and the nations of the so-called Third World that
were once European colonies.

11 Paul Hawken, *Blessed Unrest: How the Largest Movement in the World Came into Being
and Why No One Saw It Coming* (New York: Viking Penguin, 2007).

Chapter 1: Wikiiš ča?miiḥta

1 "Alaskan Village Threatened by Warming: Climate Change Is Degrading Community's
Critical Resource, Ice," ABC News, *Nightline,* 21 March 2007.

2 Throughout this book, the use of the word "traditional" refers to how my ancestors may
have described any of their activities. In its most complete sense, "traditional" refers to
an activity that is consciously embedded within a worldview supported by story, the vision quest, various teachings, and a dependable knowledge base for survival purposes.
This means that, as circumstances and conditions change over time, specific activities

also change, even though the set of assumptions underlying them do not. Consequently, when conditions change radically and forcefully, as they did during colonization, traditional activities can become so rudimentary that the word "traditional" may not seem to apply. Nevertheless, as indigenous cultures revive, as they have been doing for some decades now, indigenous peoples have a right to describe their contemporary activities as "traditional."

3 It is customarily assumed that only human beings can be creative. A Nuu-chah-nulth interpretation of "creative" assumes creativity to be, potentially, a process shared between and among all life forms.

4 A twentieth-century version of the *ƛuukʷaana* is described in Atleo, *Tsawalk: A Nuu-chah-nulth Worldview* (Vancouver: UBC Press, 2004), 99-116. It is a ceremony to remind families that, without vigilance (i.e., without remembering their grandparents' ancient teachings) their children will be lost, their families destroyed.

5 Elisabet Sahtouris, *Earthdance: Living Systems in Evolution* (Lincoln, NE: iUniverse, 2000). Elisabet's rationale for the necessity of balance and harmony is based on natural law: "Every living creature must get materials and energy from its environment to form itself and to keep itself alive ... This is why no living creature can ever be entirely independent" (51). And again: "Now we see that we are natural creatures which evolved inside a great Earthlife system. Whatever we do that is not good for life, the rest of the system will try to undo or balance in any way it can. This is why we must learn Gaia's dance and follow its rhythms and harmonies in our own lives" (67).

6 The personal spiritual power demonstrated through *tupati* is acquired through the *ʔuusumč*. In my book, *Tsawalk: A Nuu-chah-nulth Worldview,* I explain that my great-grandfather Keesta's *tupati* was his ability to throw an eagle feather at a target several metres away. A challenge would be made to anyone in the feast hall to come forward and hit the target with a feather thrown tip-first. Since a feather has a natural curve that makes it unstable in flight, and since to throw a feather tip-first creates more instability, the likelihood of hitting any target is virtually nil. Yet Keesta would hit the target every time.

7 In British Columbia, I conducted a provincewide education study that was meant to respond to a nationwide study edited by Harry Hawthorn entitled *A Survey of the Contemporary Indians of Canada,* 2 vols. (Ottawa: Indian Affairs Branch, 1966-67), better known as the Hawthorn Report. One of the Hawthorn Report's significant findings was that most Indian children did not express any hope for a viable future for themselves. My own study found that, in general, First Nations children expressed great hopes for a viable future. For example, one Grade 4 girl could not decide whether she wanted to be the prime minister of Canada or a Supreme Court judge. The difference between Hawthorn's study and my study is that conditions changed from being extremely socio-politically oppressive to being less so. In 1972, the National Indian Brotherhood submitted an education proposal entitled *Indian Control of Indian Education* (Ottawa: National Indian Brotherhood, 1972), which, in early 1973, was accepted in principle by the federal government. And, in 1982, Canada had enacted the Charter of Rights and Freedoms, which provides greater avenues of freedom for all citizens, including

members of the Aboriginal population. As Canadian society became more open and less oppressive, the First Nations children responded with expressions of hope for their own futures. See my *An Examination of Native Education in British Columbia: Kindergarten Readiness and Self-Image and Academic Achievement of the Grades 4 to 12* (Vancouver: Native Brotherhood of British Columbia, 1993).

8 BBC News, 28 July 2004, 16:03:31 GMT.

9 David Healy, *Let Them Eat Prozac: The Unhealthy Relationship between the Pharmaceutical Industry and Depression* (Toronto: James Lorimer, 2003), 402-3.

10 See "Corporate Influence on Science and Technology," presentation given by Stuart Parkinson at Green Party Spring Conference, Brighton, UK, 13 March 2004, http://www.sgr.org.uk/. A feature of the World Wide Web is that it is dynamic, it is not static or frozen in time the way library books tend to be. The downside of this is that it may appear unreliable as a scholarly source, but that is only from the cultural perspective of fixedness, which is a characteristic of literacy. The upside of the dynamic of the web is that, if a specific URL, such as the one above, does not produce the site of information, any similar alternative input will produce many like sites of information that will corroborate the citation. For example, I googled "global responsibility" just now and received 38,000,000 hits.

11 Al Gore, *The Assault on Reason* (New York: Penguin, 2007), 201.

12 Paul Hawken, *Blessed Unrest: How the Largest Movement in the World Came into Being and Why No One Saw It Coming* (Toronto: Penguin Group, 2007). This book was recommended to me by Dr. Pamela Steiner, Senior Fellow, Harvard Humanitarian Initiative, Harvard University. She has worked to bring the grassroots Israeli and Palestinian peoples together and is consequently part of the global movement identified by Paul Hawken.

13 Ibid., 11-12.

14 See David C. Korten, *The Post-Corporate World: Life after Capitalism* (San Francisco/West Harford: Berret-Koehler/Kumarian Press, 1999). For the interview with the *Sun* (magazine), see Arnie Cooper, "Everybody Wants to Rule the World: David Korten on Putting an End to Global Competition," September 2007, http://www.thesunmagazine.org/.

15 James Howard Kunstler, *The Long Emergency: Surviving the End of Oil, Climate Change, and Other Converging Catastrophes of the Twenty-First Century* (New York: Grove Press, 2006), 187.

16 Ibid., 27.

17 David C. Korten, *The Great Turning: From Empire to Earth Community* (San Francisco: Berret-Koehler, 2001) 132.

18 Kunstler, *Long Emergency*, 42, quotes the following about the limits of oil in the United States: "When a geologist named M. King Hubbert announced in 1949 that there was, in fact, a set geological limit to the supply of oil that could be described mathematically, and that it didn't lie that far in the future, nobody wanted to believe him ... By the mid-1950s, as chief of research for Shell Oil, Hubbert had worked up a series of mathematical models based on known U.S. oil reserves, typical rates of production, and apparent

rates of consumption, and, in 1956, he concluded that the oil production in the United States would peak sometime between 1966 and 1972 ... U.S. oil production proceeded to peak in 1970."

19 Fred Bunnell and Richard Atleo, *The Scientific Panel for Sustainable Forest Practices in Clayoquot Sound* (Victoria, BC: Cortex Consultants, 1995).

20 Frank J. Tipler, *The Physics of Immortality* (New York: Anchor Books, 1995), x.

21 James Lovelace, *The Vanishing Face of Gaia: A Final Warning* (London: Penguin, 2009), 40.

22 Ibid, 28-29.

23 Samuel P. Huntington, *The Clash of Civilizations and the Remaking of World Order* (New York: Simon and Schuster, 1996), 30.

24 Jean Shinoda Bolen, *Ring of Power: The Abandoned Child, the Authoritarian Father, and the Disempowered Feminine* (New York: HarperCollins, 1992), 1-2.

25 Thomas Berry, *The Dream of the Earth* (San Francisco: Sierra Club Books, 1990), 123.

26 Ibid.

27 Fritjof Capra, *The Tao of Physics: An Exploration of the Parallels between Modern Physics and Eastern Mysticism* (Glasgow: Caledonian International Book Manufacturing, 1991), 357.

28 Ibid., 357-58.

29 David B. Allison, *Reading the New Nietzsche:* The Birth of Tragedy, The Gay Science, Thus Spoke Zarathustra, *and* On the Genealogy of Morals (Lanham, MD: Rowman and Littlefield, 2001), 90.

30 Walter Kaufmann, *The Portable Nietzsche: Selected and Translated, with an Intro-duction, Prefaces, and Notes* (New York: Penguin, 1954). Kaufmann writes, in defence of Nietzsche's reputation as one who was both profound and contradictory: "Doubtless Nietzsche has attracted crackpots and villains, but perhaps the percentage is no higher than in the case of Jesus" (2).

31 Allison, *Reading the New Nietzsche,* 90.

32 Ibid., 92 [emphasis in original].

33 Erich Heller, *The Importance of Nietzsche: Ten Essays* (Chicago: University of Chicago Press, 1988), 3.

34 Kaufmann, *Portable Nietzsche,* 627 [emphasis in original].

35 Clement C.J. Webb, *A History of Philosophy* (Toronto: Oxford University Press, 1959), 9.

36 Riane Eisler, *The Chalice and the Blade: Our History, Our Future* (New York: HarperCollins, 1987), xiii, xiv.

37 Riane Eisler, *The Real Wealth of Nations* (San Francisco: Berrett-Koehler, 2007), 9 [emphasis in original].

38 Patricia Marchak and Jerry Franklin, "Foreword," in *The Rain Forests of Home: Profile of a North American Bioregion,* ed. Peter K. Schoonmaker, Bettina von Hagen, and Edward C. Wolf (Washington, DC: Island Press, 1997), x, xi.

39 Ibid.

40 Richard A. Posner, *Law, Pragmatism, and Democracy* (Cambridge, MA: Harvard University Press, 2003), 5, 31.

41 Jonathan Porritt, *Capitalism: As if the World Matters* (London: Earthscan, 2005), 80.

42 Berry, *The Dream of the Earth.*

43 Francis Fukuyama, *The End of History and the Last Man* (New York: Free Press, 2006), xviii.

44 Ibid., 45.

45 Ibid., 41.

46 Noam Chomsky, *Failed States: The Abuse of Power and the Assault on Democracy* (New York: Metropolitan Books, 2006), 38.

47 Ibid., 251.

48 Samuel P. Huntington, *Who Are We? The Challenges to America's National Identity* (New York: Simon and Schuster, 2004), 365.

49 Anthony J. Hall, *The American Empire and the Fourth World: The Bowl with One Spoon, Part One* (Montreal and Kingston: McGill-Queen's University Press, 2003), xii-xiii.

50 Ibid., xiii-xiv.

51 Peter Gay, ed., *The Enlightenment: A Comprehensive Anthology* (New York: Simon and Schuster, 1973), 384.

52 Immanuel Kant, "On History," in Gay, *The Enlightenment,* 805.

53 Chellis Glendinning, *My Name Is Chellis and I'm in Recovery from Western Civilization* (Boston/New York: Shambhala Theory/Vintage Books, 1994), 61.

54 Thomas Lewis, Fari Amini, and Richard Lannon, *A General Theory of Love* (New York: Random House, 1978), 68-69.

55 Ibid., 85.

56 George Soros, *The Age of Fallibility: Consequences of the War on Terror* (New York: Public Affairs, 2006), xxv.

57 Ibid., 14.

58 Gilbert M. Sproat, *Nootka: Scenes and Studies of Savage Life* (1968; repr., Victoria, BC: Sono Nis Press, 1987), 3-4.

59 Although the question "why," when asked from the position of a search for knowledge and truth, is an effective form of scientific endeavour, it can also be symptomatic of separation from the spiritual, especially when the question is about poverty, suffering, disease, and natural disasters such as earthquakes and tsunamis.

60 Brian Greene, *The Elegant Universe: Superstrings, Hidden Dimensions, and the Quest for the Ultimate Theory* (New York: Vintage Books, 1999), 169.

Chapter 2: Mirrors and Patterns

1 Carobeth Laird, *Mirror and Pattern: George Laird's World of Chemehuevi Mythology* (Banning, CA: Malki Museum Press, 1984), 17.

2 The story of my great-grandfather, Keesta, who grew into adulthood after the manner of his ancestors, is told in Atleo, *Tsawalk: A Nuu-chah-nulth Worldview* (Vancouver: UBC Press, 2004), 77-81. Robin Fisher, *Contact and Conflict* (Vancouver: UBC Press, 1986), xiv, xv, 48, explains that the fur trade had a minimally disruptive effect on coastal cultures since the Aboriginals commanded sovereign control over the trading process.

Trade was a common local practice, as the newcomers from Europe were to discover. Consequently, the Nuu-chah-nulth culture remained intact during the period when Keesta grew up. The significance of this recent cultural history is that the argument made for the cultural integrity of Keesta's day is also made for the formative years of my life.

3 On the issue of Western influence on indigenous peoples, see Rolf Knight, *Indians at Work: An Informal History of Native Indian Labour in British Columbia: 1858-1930* (Vancouver: New Star Books, 1978). Knight's intention in this work seems to have been to encourage the notion that indigenous peoples had already successfully integrated into Western society by 1930, at least in so far as work was concerned. His arguments below appear persuasive, but, significantly, they arise from an etic perspective. Knight's main argument is based on Claude Lévi-Strauss's notion that indigenous societies, in order to remain pure, cannot change in any way. According to Knight: "The original ethnographic accounts of Indian societies were primarily reconstructions of conditions which existed in the mid or later half of the nineteenth century. They largely were reminiscences of societies which existed after three generations of fur trade and contact with Europeans had effected [sic] native societies and economies. During that period new factors altered features of native societies" (32). Knight continues: "In sum then, by 1858, at the beginning of massive European settlement, no truly indigenous and unchanged Indian society remained in BC. They had all undergone variable but considerable degrees of change. They were neo-traditional Indian societies undergoing further change. Much of what is today popularly held to be the original, the pre-contact cultures of Indian peoples were partly the formations of the eighty year fur trade history. There were both continuities with past indigenous cultures and newly emergent forms" (233).

4 Willis W. Harman and Elisabet Sahtouris, *Biology Revisioned* (Berkeley, CA: North Atlantic Books, 1998), 142.

5 The Tla-o-qui-aht are a Nuu-chah-nulth community in Clayoquot Sound. They are situated in two places, on a reserve across the local waterway from Tofino and on a reserve located at one end of Long Beach.

6 Precontact Nuu-chah-nulth had large wooden homes, made from cedar, that were designed for extended families. The advent of European settlement introduced the smaller nuclear homes, and, consequently, the traditional large wooden homes became known in English as "big houses." Since each large house had a name or several names during ancient times, it is still appropriate to reference oneself as belonging to a particular big house (Keesta's big house had three names, two of which referred to oil and one that referred to cedar).

7 Linda Tucker, *Mystery of the White Lions: Children of the Sun God* (Mapumulanga, South Africa: Npenvu Press, 2003), 55. Tucker quotes Mutwa, reputedly one of the foremost shamans of South Africa, as follows: "Much of the violence and stupid activity in our world is because people are still enslaved to this thing called fear and have not come to see the connection between all things" (54).

8 Ibid., 52. Mutwa explained to Tucker how his grandfather began to teach him, a little bit at a time, that the spirit of the Star Gods lives in everything. His grandfather showed him

a tree and asked him what it was. "A tree, grandfather," replied Mutwa. His grandfather said, "This is not a tree, you little dog, this is a person" (53).

9 Laird, *Mirror and Pattern,* 17 [emphasis added].

10 Samuel P. Huntington (1996). *The Clash of Civilizations and the Remaking of World Order* (New York: Simon and Schuster, 1996), 53.

11 Jean Shinoda Bolen, *Ring of Power: The Abandoned Child, the Authoritarian Father, and the Disempowered Feminine* (New York: HarperCollins, 1992), 1.

12 "Scientific theories" is used here in the current hegemonic sense. In the Western world, the study of scientific theory prevails within the educational system, which is the medium used to reify culture. Religion plays a role but not in mainstream education or politics.

Chapter 3: Genesis of Global Crisis

1 Samuel P. Huntington, *The Clash of Civilizations and the Remaking of World Order* (New York: Simon and Schuster, 1996), 53.

2 Ibid., 53.

3 Jack Weatherford, *Genghis Khan and the Making of the Modern World* (New York: Three Rivers Press, 2004).

4 Ibid., 236.

5 Ibid., 7 [emphasis added].

6 Ibid., xxv-xxvi.

7 Charles C. Mann, *1491: New Revelations of the Americas before Columbus* (New York: Alfred A. Knopf, 2005), 336.

8 Weatherford, *Genghis Khan,* xxiii-xxiv.

9 Ibid., xxii.

10 Ibid., xxiii.

11 Ibid., 33.

12 Ibid., 142.

13 Ibid., xxiii.

14 E. Richard Atleo (Umeek), *Tsawalk: A Nuu-chah-nulth Worldview* (Vancouver: UBC Press, 2004), 93.

15 Graham Saayman: *Hunting with the Heart: A Vision Quest to Spiritual Emergence* (Rondebosch: South Africa, 2007), 64.

16 Ibid., 65.

17 Tjalling Halbertsma, "When East Found West," *South China Morning Post,* 14 April 2001.

18 Ibid.

19 Ronald Wright, *Stolen Continents: The "New World" through Indian Eyes since 1492* (Toronto: Viking, 1992), 11-12.

20 Jean-Jacques Rousseau, *The Social Contract,* trans. Maurice Cranston (London: Penguin, 1968), 49.

21 C.J. Jaenen, "Amerindian Views of French Culture in the Seventeenth Century," in *Out of the Background: Readings on Canadian Native History,* ed. R. Fisher and K. Coates (Toronto: Copp Clark Pitman, 1987), 121.

22 Amerigo Vespucci, *The First Four Voyages of Amerigo Vespucci*, reprinted in facsimile and translated from the rare edition (Florence, 1505-06) (London: Bernard Quaritch, 1893). The following reference to Vespucci is from Sir Thomas More, *Utopia* (1516; reprint, London: Cassell and Co., 1901): "This Raphael, who from his family carries the name of Hythloday, is not ignorant of the Latin tongue, but is eminently learned in the Greek, having applied himself more particularly to that than to the former, because he had given himself much to philosophy, in which he knew that the Romans have left us nothing that is valuable, except what is to be found in Seneca and Cicero. He is a Portuguese by birth, and was so desirous of seeing the world, that he divided his estate among his brothers, ran the same hazard as Americus Vesputius, and bore a share in three of his four voyages that are now published; only he did not return with him in his last, but obtained leave of him, almost by force, that he might be one of those twenty-four who were left at the farthest place at which they touched in their last voyage to New Castile" (2). Although More's *Utopia* was not meant as a factual account of any people or place, there seems no doubt that More was one of many scholars of that time who had read the letters of Vespucci.
23 Vespucci, *First Four Voyages,* 87.
24 Ibid., 8.
25 Jack Weatherford, *Indian Givers: How the Indians of the Americas Transformed the World* (New York: Crown Publishers, 1986), 122.
26 Ibid., 124.
27 Vespucci, *First Four Voyages,* 8.
28 Weatherford, *Indian Givers,* 122.
29 Ibid., 12-13.
30 Ibid., 27.
31 Ibid., 28.
32 Atleo, *Tsawalk,* 15.
33 Vespucci, *First Four Voyages,* 17.
34 Ibid., 29-30.
35 Weatherford, *Indian Givers,* 122.
36 Sir Thomas More, *More's Utopia: A Dialogue of Comfort* (London: Dent, 1915), 15.
37 Ibid., 104 [emphasis added].
38 Vespucci, *First Four Voyages,* 10.
39 Weatherford, *Genghis Khan,* 12.
40 Ibid., 33.

Chapter 4: The Nuu-chah-nulth Principle of Recognition

1 Jeremy Narby, *Intelligence in Nature: An Inquiry into Knowledge* (New York: Penguin, 2006), 1.
2 Ibid., 150-51.
3 Ibid., 35-36.

4 See Charles I. Bevans, comp., *Treaties and Other International Agreements of the United States of America, 1776-1949*, vol. 3, *Multilateral, 1931-1945* (Washington, DC: Department of State, 1969), art. 1.

5 Ibid.

6 Personal Communication with Shawn A-in-chat Atleo, National Chief, Assembly of First Nations, Canada, 7 December 2009.

7 Section 35 reads:

> 35. (1) The existing aboriginal and treaty rights of the aboriginal peoples of Canada are hereby recognized and affirmed.
>
> (2) In this Act, "aboriginal peoples of Canada" includes the Indian, Inuit and Métis peoples of Canada.
>
> (3) For greater certainty, in subsection (1) "treaty rights" includes rights that now exist by way of land claims agreements or may be so acquired.
>
> (4) Notwithstanding any other provision of this Act, the aboriginal and treaty rights referred to in subsection (1) are guaranteed equally to male and female persons.

8 Thomas Berger, *A Long and Terrible Shadow: White Values, Native Rights in the Americas: 1492-1992* (Vancouver: Douglas and McIntyre, 1991), 150-51.

9 Harold Cardinal, *The Unjust Society: The Tragedy of Canada's Indians* (Edmonton: Hurtig, 1969).

10 *Delgamuukw* was an Aboriginal land claim case heard in the Supreme Court of British Columbia and recorded in *Reason for Judgment: of the Honourable Chief Justice Allan McEachern*, 8 March 1991, Smithers Registry No. 0843. In general, McEachern, in denying the legal credibility of the oral histories presented by the plaintiffs, "could make no other decision than that which [he] made." Or so he told me at Walter Koerner's ninety-fifth birthday party in Vancouver. In 1997, the Supreme Court of Canada reversed this decision concerning the admissibility of Aboriginal oral histories in courts of law and, moreover, established the constitutional recognition for Aboriginal rights and title. To this day, however, controversy surrounds the interpretation of these rights.

11 Thomas Berry, *The Dream of the Earth* (San Francisco: Sierra Club Books, 1990), 123.

12 Elisabet Sahtouris, *Earthdance: Living Systems in Evolution* (Lincoln, NE: iUniverse, 2000), 30.

13 C.G. Jung, *Modern Man in Search of a Soul,* trans. W.S. Dell and Cary F. Baynes (New York: A Harvest/HBJ Book, 1933), 181.

14 Ibid., 119.

15 Sixty-first session of the UN General Assembly, "Agenda Item 68: Report of the Human Rights Council," 7 September 2007.

Chapter 5: The Nuu-chah-nulth Principle of Continuity

1 Section 1 of the Charter states: "The Canadian Charter of Rights and Freedoms guarantees the rights and freedoms set out in it subject only to such reasonable limits prescribed by law as can be demonstrably justified in a free and democratic society."

2 Ahous is a precontact Nuu-chah-nulth place name. With the publication of *The Nootka: Scenes and Studies of Savage Life* (Victoria: Sono Nis Press, 1987) in 1868, Gilbert Sproat suggested that the Nuu-chah-nulth people should be called "Aht," a sound he often heard. But "aht" means "person of" or "person from," so that an Ahous-aht is a person "from Ahous." Over time, Ahous-aht, like all other Nuu-chah-nulth communities (e.g., Tla-o-qui-aht, Kyuquot, Dididaht, etc.), were recorded as place names by the Government of Canada.

3 Thomas Lewis, Fari Amini, and Richard Lannon, in *A General Theory of Love* (New York: Random House, 1978), note that "human physiology is (at least in part) an open-loop arrangement, an individual does not direct all of his own functions. A second person transmits regulatory information that can alter hormone levels, cardiovascular function, sleep rhythms, immune function, and more – inside the body of the first. The reciprocal process occurs simultaneously: the first person regulates the physiology of the second, even as he himself is regulated. Neither is a functioning whole on his own; each has open loops that only somebody else can complete. Together they create a stable, properly balanced pair of organisms ... A baby's physiology is maximally open-loop: without limbic regulation, his vital rhythms collapse and he will die" (85).

4 Elisabet Sahtouris, *Earthdance: Living Systems in Evolution* (Lincoln, NE: iUniverse, 2000), 87-103.

5 Carobeth Laird, *Mirror and Pattern: George Laird's World of Chemehuevi Mythology* (Banning, CA: Malki Museum Press, 1984), 17.

6 These stories are found in Atleo, *Tsawalk: A Nuu-chah-nulth Worldview* (Vancouver: UBC Press, 2004).

7 Sea urchins are a favourite traditional Nuu-chah-nulth food. Sea urchins contain medicinal properties that prevent infections from abrasions and cuts received during the cleaning of fish and other foods.

8 Michael Asch, *Home and Native Land: Aboriginal Rights and the Canadian Constitution* (Toronto: Methuen, 1984), 112.

9 "An Act Further to Amend the Indian Act, 1880," chapter 27, assented to 19 April 1884, in Canada, *Indian Acts and Amendments, 1868-1950* (Ottawa: Indian and Northern Affairs Canada, 1981), 52.

10 Quoted in Charles Hou, *To Potlatch or Not to Potlatch: An In-depth Study of Culture Conflict between the BC Coastal Indian and the White Man* (Vancouver: British Columbia Teachers' Federation, n.d.), 9.

11 After rescuing the children of Ahous, who had been stolen by Pitch Woman, Son of Mucus went back to his spiritual home to marry the daughter of the chief. Later, he returned to earth to create biodiversity by transforming all life forms, which, at this time, were of one species.

12 Indian Act, R.S.C. 1906, chapter 81, part 1, "Indians: Schools," in Canada, *Indian Acts and Amendments, 1868-1950* (Ottawa: Indian and Northern Affairs Canada, 1981), 101-2.

13 Our traditional precontact form of governance included hereditary chiefs, advisors, councillors, messengers, speakers, historians, and a number of roles that might parallel the portfolio system of contemporary governance. The same people who had served in

the precontact form of governance continued to serve as though they had been elected under the Indian Act.

14 Brian E. Titley, *A Narrow Vision: Duncan Campbell Scott and the Administration of Indian Affairs in Canada* (Vancouver: UBC Press, 1986), 163.

15 Historically, there has tended to be a difference between the way Europeans and Aboriginals practise spirituality. Some highly spiritual Europeans are monastic in their tendency to isolate themselves from society. Aboriginals isolate themselves during *ʔuusumč* but then bring back the substance of their quests so that it may be used within their communities.

16 Philip Drucker, *Northern and Central Nootkan Tribes* (Washington, DC: United States Government Printing Office, 1951), 11-12.

17 Sproat, *The Nootka*, 132-33.

18 Ibid., 135.

19 Quoted in C.J. Jaenen, "Education for Francization: The Case of New France in the Seventeenth Century," in *Indian Education in Canada*, vol. 1, *The Legacy*, ed. J. Barman, Y.M. Hebert, and D. McCaskill (Vancouver: UBC Press, 1986), 58.

20 James Tully, *Strange Multiplicity: Constitutionalism in an Age of Diversity* (New York: Cambridge University Press, 1995), James Tully, "Reconciling Struggles over the Recognition of Minorities: Towards a Dialogical Approach," in *Diversity and Equality: The Changing Framework of Freedom in Canada*, ed. Avigail Eisenberg (Vancouver: UBC Press, 2006).

21 Reasons for Judgment, the Honourable Chief Justice Allan McEachern, Supreme Court of British Columbia, *Delgamuukw v. British Columbia*, file no. 0834, Smithers Registry, p. 14.

22 Quoted in Peter Laslett, *John Locke: Two Treatises of Government – A Critical Edition with an Introduction and Apparatus Criticus* (London: Cambridge University Press, 1967), 317 [emphasis in original].

23 Ibid., 309 [emphasis in original].

24 H.B. Hawthorn, C.S. Belshaw, and S.M. Jamieson, *The Indians of British Columbia: A Study of Contemporary Social Adjustment* (Toronto: University of Toronto Press, 1958), 70.

Chapter 6: The Nuu-chah-nulth Principle of Continuity

1 In Atleo, *Tsawalk: A Nuu-chah-nulth Worldview* (Vancouver: UBC Press, 2004), I define *iisʔak̓*, pronounced "ee-sock," as "sacred respect." The phrase is an attempt to indicate the unity of the physical (respect that ascribes value to life) with the spiritual (sacred, the source of life). *Iisʔak̓*, when applied to reality, means that everything has value and therefore is not to be treated without regard. In a polarized reality, value must be either positive or negative.

2 The same argument can be made about the writing of Charles Darwin and other scholars from the Ages of Reason and Enlightenment.

3 My use of the article "a" is vital here. The principle of integrity of being, so important to human development, is also vital to the notion that each has his or her own way of life. The outward manifestations of "way of life" – as in preferences for specific foods, types of clothing, or items of ritual – are all variations on a common theme.

4 Elisabet Sahtouris, *Earthdance: Living Systems in Evolution* (Lincoln, NE: iUniverse, 2000). Sahtouris affirms the interconnective nature of existence, of being, by analyzing the historical development of the I, or ego, and the imbalance this has created in the natural order of creation. She writes: "Our ability to be objective, to see ourselves as the I or eye of our cosmos, as beings independent of nature, has inflated our egos – ego being the Greek word for I. We came to separate the I from the it and to believe that 'it' – the world apart from us, out there – was ours to do with as we pleased. We told ourselves we were either God's favored children or the smartest and most powerful naturally-evolved creatures on Earth. This egotistic attitude has been very much a factor in bringing us to adolescent crisis ... And so an attitude of greater humility and willingness to accept some guidance from our parent planet will be an important factor in reaching our species maturity" (8). She continues: "A single notion that would account for such pattern is the concept of mutual consistency ... This is the concept that the universe is a dynamic web of events in which no part or event is fundamental to the others since each follows from all the others, the relations among them determining the entire cosmic pattern or web of events. In this conception, all possible patterns of cosmic matter-energy will form, but only those working out their consistency with surrounding patterns will last. Mutual means shared; consistency means agreement or harmony" (23).

5 Ibid., 23.

6 John Dunn, *Democracy: A History* (Toronto: Penguin, 2006), 188. About democracy and its recent emergence on the world stage, Dunn writes: "What we mean by democracy is not that we govern ourselves. When we speak or think of ourselves as living in a democracy, what we have in mind is something quite different. It is that our own state, and the government which does so much to organize our lives, draws its legitimacy from us, and that we have a reasonable chance of being able to compel each of them to continue to do so. They draw it, today, from holding regular elections, in which every adult citizen can vote freely and without fear, in which their votes have at least a reasonable equal weight, and in which any uncriminalized political opinion can compete freely for them" (19-20).

7 Ibid., 15.

8 Ibid., 17 [emphasis in original].

9 Francis Fukuyama, *The End of History and the Last Man* (New York: Free Press, 2006), 45 [emphasis in original].

10 In his book, *The End of History*, Fukuyma notes the following distinctions between liberalism and democracy:

Liberalism and democracy, while closely related, are separate concepts. Political liberalism can be defined simply as a rule of law that recognizes certain individual rights or freedoms from government control. While there can be a wide variety of definitions of fundamental rights, we will use the one contained in Lord Bryce's classic work on democracy, which limits them to three: civil rights, "the exemption from control of the citizen in respect of his person and property"; religious rights, "exemption from control in the expression of religious opinions and the practise of worship"; and what he calls political rights, "exemption from control in matters which do not so plainly affect the welfare of the whole community as to render control necessary," including the fundamental right of press freedom ... Democracy, on the other hand, is the right held universally by all citizens to have a share of political power, that is, the right of all citizens to vote and participate in politics. (42-43)

He continues:

It is possible for a country to be liberal without being particularly democratic, as was eighteenth-century Britain. A broad list of rights, including the franchise, was fully protected for a narrow social elite, but denied to others. It is also possible for a country to be democratic without being liberal ... A good example of this is the contemporary Islamic Republic of Iran, which has held regular elections that were reasonably fair by Third World standards, making the country more democratic than it was in the time of the Shah. Islamic Iran, however, is not a liberal state; there are no guarantees of free speech, assembly, and, above all, of religion. (43-44)

11 Chief Justice H.E. Hutcheon, *Delgamuukw v. British Columbia,* file no. 23799, 16 and 17 June 1997, 11 December 1997, paras. 127 and 186.

12 Jack Weatherford, *Native Roots: How the Indians Enriched America* (New York: Fawcett Columbine, 1991), 17-18.

13 Thomas Berger, *A Long and Terrible Shadow: White Values, Native Rights in the Americas, 1492-1992* (Vancouver: Douglas and McIntyre), 66-67.

14 For more information about Walter Koerner, see "Biography of Walter C. Koerner," http://www.library.ubc.ca/koerner/celebration/wkoerner.html.

15 A.C. Hamilton, *A Feather Not a Gavel: Working towards Aboriginal Justice* (Winnipeg: Great Plains, 2001), 20.

16 Douglas Cole, *Captured Heritage: The Scramble for Northwest Coast Artifacts* (Seattle: University of Washington Press, 1985), 200.

17 Ibid., 201.

18 Michael J. Chandler, Christopher E. Lalonde, Bryan W. Sokol, and Darcy Hallett, "Surviving Time: Aboriginality, Suicidality, and the Persistence of Identity in the Face of Radical Developmental and Cultural Change" (draft paper, 2003), 3.

19 Ibid.

20 John Ralston Saul, *A Fair Country: Telling Truths about Canada* (Toronto: Penguin, 2008), 24.

21 Ibid., 3.

22 *Delgamuukw v. British Columbia,* [1997] 3 S.C.R. 1010, Docket 23799, 11 December 1997, para. 150.

23 Neil Vallance, "Misuse of 'Culture' by the Supreme Court of Canada," in *Diversity and Equality: The Changing Framework of Freedom in Canada,* ed. Avigail Eisenberg (Vancouver: UBC Press, 2006), 109.

24 Robin Fisher, *Contact and Conflict: Indian-European Relations in British Columbia, 1774-1890* (Vancouver: UBC Press, 1977), 48.

25 Charles C. Mann, *1491: New Revelations of the Americas before Columbus* (New York: Alfred A. Knopf, 2005), 16. Mann's book is a comprehensive and well-researched account of the Americas before contact. It is a scholarly piece that is self-conscious about the ways that European definitions of words like "civilization," "tribe," "chief," "king," and "emperor" distort perceptions of both European and Aboriginal societies, extolling the image of the former and denigrating the image of the latter. This book is one good example of a more balanced view of precontact societies – one that could lead to greater mutual understanding and, I would hope, to mutual respect.

26 Ibid., 250, 251.

27 José Mariano Moziño, *Noticias de Nutka: An Account of Nootka Sound 1792* (Vancouver: Douglas and McIntyre, 1970), 24.

28 Ibid., 9 [emphasis added].

29 Amerigo Vespucci, *The First Four Voyages of Amerigo Vespucci,* reprinted in facsimile and translated from the rare edition (Florence, 1505-06) (London: Bernard Quaritch, 1893), 8.

30 Quoted in C.J. Jaenen, "Education for Francization: The Case of New France in the Seventeenth Century," in *Indian Education in Canada,* vol. 1, *The Legacy,* ed. J. Barman, Y.M. Hebert, and D. McCaskill (Vancouver: UBC Press, 1986), 58.

Chapter 7: Haḥuułism

1 Words spoken by Mary Little, daughter of Manhousaht Ḥaw'ił.

2 Thomas Berry, *The Dream of the Earth* (San Francisco: Sierra Club Books, 1990), 130.

3 Fritjof Capra, *The Tao of Physics: An Exploration of the Parallels between Modern Physics and Eastern Mysticism* (Glasgow: Caledonian International Book Manufacturing, 1991), 357.

4 Patricia Marchak and Jerry Franklin, "Foreword," in *The Rain Forests of Home: Profile of a North American Bioregion,* ed. Peter K. Schoonmaker, Bettina von Hagen, and Edward C. Wolf (Washington, DC: Island Press, 1997), x.

5 *Qua* is a Nuu-chah-nulth word that I think is more suited to describing the nature of *Haḥuułism* than is the word "philosophy." Philosophy employs the use of reason and argument in seeking knowledge about reality. Although *qua* also employs reason and argument in its search for the nature of reality, its usage in Nuu-chah-nulth is supported by the knowledge acquisition method of the *ʔuusumč* as well as the sacred ceremony of the *łuukʷaana,* which contains in its middle syllable the essence of the meaning of the creator, Ḱʷaaʔuuc. In this sense, *qua* is not a theory, opinion, or point of view based solely on reason and logic but, rather, includes in its meaning the collective experience of Nuu-chah-nulth practitioners of the vision quest.

6 Evalyn Gates, *Einstein's Telescope: The Hunt for Dark Matter and Dark Energy in the Universe* (New York: W.W. Norton, 2009), 26.

7 Jeremy Narby, *Intelligence in Nature: An Inquiry into Knowledge* (New York: Penguin, 2006), 1.

8 This story of the Beaver people is recounted in Hugh Brody, *Maps and Dreams: Indians and the British Columbia Frontier* (Markham, ON: Penguin Books, 1981), 45.

9 Ida Halpern and David Duke, "'A Very Agreeable Harmony': Impressions of Nootkan Music," *Sound Heritage* 7, 1 (1978): 63.

10 Ibid., 63.

11 George Clutesi, *Potlatch* (Sidney, BC: Gray's Publishing, 1969). George introduces his eyewitness account as follows: "This narrative is not meant to be documentary. In fact it is meant to evade documents. It is meant for the reader to feel and to say I was there and indeed I saw" (5). What is remarkable here is that Clutesi attempts to place Nuu-chah-nulth ceremony within its proper context, which is dynamic and immediate, rather than into the frozen context of a documentary, which is both dead and historical.

12 Ibid., 21.

13 Ibid., 21-22.

14 Ibid., 24.

15 Ibid.

16 Metaphorically, it can be said that all babies take whatever they want, that is, they simply steal. However, babies are never accused of stealing because they are immature, they are babies. In the beginning, until they learn otherwise, both Son of Raven and Bear simply try to take what they want.

17 Amerigo Vespucci, *The First Four Voyages of Amerigo Vespucci*, reprinted in facsimile and translated from the rare edition (Florence, 1505-06) (London: Bernard Quaritch, 1893), 8.

18 The concept of a "beginning" can only be grasped from a human perspective, which is defined by natural limitations; therefore, it is not possible for any human to witness an actual beginning to creation.

19 Although there remain many within the academy that continue to question the validity of indigenous knowledge, there are others, such as John Ralston Saul, Wade Davis, Jeremy Narby, and Elisabet Sahtouris who accept indigenous knowledge as another way of knowing.

20 Elisabet Sahtouris, *Earthdance: Living Systems in Evolution* (Lincoln, NE: iUniverse, 2000), 26.

21 The notion of hereditary chieftainship has been greatly distorted by misleading parallels to European monarchy. All leaders received extensive training that began at birth, with the result that their potential for leadership became evident long before they came of age. In addition, every chief had a set of elder counsellors to help with governance. With regard to gender, which is so important in the Western world, among the ancient Nuu-chah-nulth men and women had equal access to the means of power and influence, which was the vision quest. Many Nuu-chah-nulth have declared that female shamans

were more celebrated, more powerful and influential, than were male shamans. A chief-tainship was one seat of power and influence, and it was held in balance by the power and influence of those with shamanic gifts.

22 John Dunn, *Democracy: A History* (Toronto: Penguin, 2005), 188.

23 E.R. Atleo, *Tsawalk: A Nuu-chah-nulth Worldview* (Vancouver: UBC Press, 2004), 68.

24 Charles Darwin, *On the Origin of Species by Means of Natural Selection, or the Preservation of Favoured Races in the Struggle for Life* (London: John Murray, 1859).

Index

Printed and bound in Canada

Set in Garamond and Times New Roman by Artegraphica Design Co. Ltd.

Text design: Irma Rodriguez

Copy editor: Joanne Richardson

Proofreader: Frank Chow

Indexer: Lillian Ashworth